Springer Series in Optical Sciences   Volume 28

Edited by David L. MacAdam

Springer Series in Optical Sciences

Edited by David L. MacAdam

Editorial Board: J. M. Enoch   D. L. MacAdam   A. L. Schawlow   T. Tamir

Orestes N. Stavroudis

Modular
Optical Design

With 54 Figures

Springer-Verlag Berlin Heidelberg GmbH 1982

Professor ORESTES N. STAVROUDIS
Optical Sciences Center, University of Arizona
Tucson, AZ 85721, USA

ISBN 978-3-662-14473-2   ISBN 978-3-540-38801-2 (eBook)
DOI 10.1007/978-3-540-38801-2

Library of Congress Cataloging in Publication Data. Stavroudis, O. N. (Orestes Nicholas), 1923- Modular optical design. (Springer series in optical sciences ; v. 28) Includes bibliographical references and index. 1. Optical instruments–Design and construction. I. Title. II. Series. QC372.2.D4S7 681'.4 81-14336 AACR2

Originally published by Springer-Verlag Berlin Heidelberg New York in 1982.
Softcover reprint of the hardcover 1st edition 1982

Offset printing: Beltz Offsetdruck, Hemsbach/Bergstr.

2153/3130-543210

Quando che 'l cubo con le cose appresso,
       Se agguaglia à qualche numero discreto:
       Trouan dui altri, differenti in esso.
Dapoi Terrai, questo per consueto,
       Che 'l lor produtto, sempre sia equale
       Al terzo cubo, delle cose neto
El residuo poi suo generale,
       Delli lor lati cubi, ben sostratti
       Varra la tua cosa principale.
In el secondo, de cotesti atti;
       Quando che 'l cubo restasse lui solo,
       Tu asseruerai quest' altri contratti,
Del numer farai due, tal part' à nolo
       Che l'una, in l'altra, si produca schietto,
       El terzo cubo delle cose in stolo
Delle quai poi per commun precetto,
       Torrai li lati cubi, insieme gionti
       Et cotal somma, sarà il tuo concetto:
El terzo, poi de questi nostri conti,
       Se solue col secondo, se ben guardi
       Che per natura son quali congionti
Questi trousi, & non con passi tardi
       Nel mille cinquecent' è quattro è trenta;
       Con fondamenti ben saldi, e gagliardi,
Nella Citta dal mar' intorno centa.

TARTAGLIA--1539

Foreword

Images are ubiquitous. Their formation is one of natures universalities. Water droplets in suspension act in concert to produce rainbows. A partially filled wine glass can be made to form the image of a chandelier at a boring dinner party. The bottom of a water glass, too, can be made to produce an optical image, wildly distorted perhaps, but nevertheless recognizable as an optical image.

Primitive folklore abounds with images. Perseus used his highly polished shield as a rear view mirror to lop off Medusa's head without turning himself into stone. Narcissus, displaying incredibly poor taste, fell in love with his own reflection in a pool of water, causing poor Echo to pine away to a mere echo and providing yet another term for the psychoanalytic lexicon.

Strepsiades, according to Aristophanes, proposed using a "burning stone" to melt a summons off the bailiff's wax tablet. And the castaways in Jules Vernes' *Mysterious Island* made a burning glass by freezing water in a watch crystal. Everyone from the Baron Münchhausen to Tom Swift has gotten into the optics act with incredible but eminently useful optical devices.

Indeed, Mother Nature herself has had a hand in evolving image-making devices. Any reasonably symmetric glob of transparent material, such as an aggregate of cells, is capable of forming an image. It is not difficult to imagine the specialization of such an aggregate into a blastula-like structure with an anterior window and light sensitive neurons at its posterior region. That such an organ would evolve into an eye, able to form and perceive images of an external universe, would appear to be in inevitable consequence of the pressures of natural selection. It is therefore not coincidence that the higher Mollusks and the higher Chordates, two phila vastly dissimilar biochemically, morphologically, and environmentally, have evolved eyes which are almost identical structurally.

Perhaps one can dare to say that optical design is *the* most ancient profession although not necessarily as disreputable as another that shares this distinction. There has always been something of the arcane about it. The lens grinder's laboratory, like the alchemist's, was a scary place. The men that labored in them also must have appeared strange and remote, like Baruch de Spinoza, banished and excommunicated, who made his living as an optician to support his philosophical writings.

In this day and age, lens design is still more an art than a science. In spite of the use of computers and computer programs in optical design, its skills are best learned by practical exercise, by experience, by doing — not by learned treatises such as this. In this sense these skills, to the outsider, smell of the occult. One suspects lens designers of casting spells and uttering incantations to master, if not to exorcise, the evil aberrations. This sort of wzardry goes beyond my puny powers. All I can do is provide the book. The bell and candle I must leave to others.

Preface

This book offers no spells, no incantations, no whiff of incense, not even
something as practical as a ready-made ju-ju. It is concerned with a new approach to the initial stages of optical design. It offers a collection of
concepts and the accompanying equations for generating third- and fifth-order
optical designs. When properly set up, the process is rapid and economical.
A computer is required; the backs of envelopes, slide rules, desk calculators
and abaci are all insufficiently rapid or accurate to perform the required
calculations. The process is not, on the other hand, automatic design. The
guidance of an experienced and wise operator is imperative.

We begin with a chapter on paraxial and third-order preliminaries, followed by a chapter on the Delano y-$\bar{y}$ diagram. This we have found useful, not
so much as a design tool, but as a heuristic device for revealing the secrets
of the module. In the third chapter, the two-surface system is studied ex-
haustively, a preliminary to the fourth chapter in which the module is defined
and some of its properties revealed. In the fifth chapter further properties
are studied in terms of the module's singularities.

The point of the proposition is reached in Chapter 6, where the method of
assembling modules into lens designs is discussed and the appropriate lens-
design equations are derived. It is here we find how to make third-order aberrations vanish. Chapter seven extends these results to the domain of the fifth-
order aberrations and of seventh-order spherical aberration. Chapter 8 contains
several examples. In Chapter 9 is found an appropriate coda, complete with the
usual speculation about future developments.

Acknowledgments. The research, the results of which are presented here, was
conducted over a period of many years. It began while I was employed in the
Optics and Metrology Division at the National Bureau of Standards in Washing-
ton, D.C. and was continued at the University of Arizona under the leadership
of Aden Meinel and Peter Franken, past Director and Director, respectively,
of the Optical Sciences Center, whose generous support and encouragement I
gratefully acknowledge. Thanks are also due to the Department of Defense's
project THEMIS, and to the Perkin Foundation whose funds made this work pos-

sible. My thanks go also to my students, Frank Powell, Romeo Mercado, and Douglas Anderson for daring to undertake extensive research in such an un-fashionable area. For her infinite patience and her tacit but palpable sup-port I thank my wife Dorle, especially for her tolerance of the persistent piles of peculiar litter that accumulated wherever I happened to be working. I must thank Keith Treptow, M.D., whose expert and wise ministrations helped extract me from a deep and dismal abyss into which I fell during the prepa-ration of this book. Finally, I thank Elena Bennett, who cheerfully and ex-pertly typed this mess.

Tucson, Arizona
June, 1981   *Orestes Stavroudis*

Contents

List of Symbols

Optical	Canonical		Optical	Canonical	
c_i	C_i	Surface curvature	b	B	Spherical aberration, image
t_i	T_i	Thickness, separation	$\bar{b}$	$\bar{B}$	, pupil
t_0	T_0	Distance, main focus to surface	$\oint$	F	Coma, image
$\bar{t}$	E	Distance, main focus to pupil	$\bar{\oint}$	$\bar{F}$	, pupil
$\bar{t}'$	A	Distance, 2nd surface to pupil	c	C	Astigmatism, image
$\bar{t}$	G	Distance, 1st surface to pupil	$\bar{c}$	$\bar{C}$	, pupil
u	U	Marginal-ray slope	p	P	Petzval
y	Y	height	e	E	Distortion, image
$\bar{u}$	$\bar{U}$	Chief-ray slope	$\bar{e}$	$\bar{E}$	, pupil
$\bar{y}$	$\bar{Y}$	height	s	S	$\Delta(u/n)$, image
a	A	Refraction invariant	$\bar{s}$	$\bar{S}$	, pupil
h	H	LaGrange invariant	t	$\mathscr{T}$	Transverse chromatic
			ℓ	$\mathscr{L}$	Axial chromatic

1. Introduction

1.1 General Background

Some years ago I was asked to design a two-mirror optical system with a variable magnification, which would focus radiation from an extended object onto
a detector. The problem was almost trivial; so much so that it appeared to
be solvable by analytic means as well as by the usual numerical techniques.
I made the attempt. Many months later — indeed, long after the original purpose of the design had been forgotten — and after an extensive excursion into
the never-never land of cubic polynomials, I had a solution. Not a solution
to the original problem, I hasten to add, but nevertheless a solution.

It was, in fact, a two-parameter family of solutions. Each member of the
family was a Cassegrain-like or a Gregorian-like system consisting of two
spherical mirrors having the unusual property that, relative to a focal plane
and its image at infinity, third-order spherical aberration was zero for
every choice of a marginal ray. In what follows, we will refer to this focal
plane as the *main focus* (Figs.1.1,2).

One of the parameters of this family of optical elements is only a scale
factor, which controls the focal length, or the power. We will refer to this

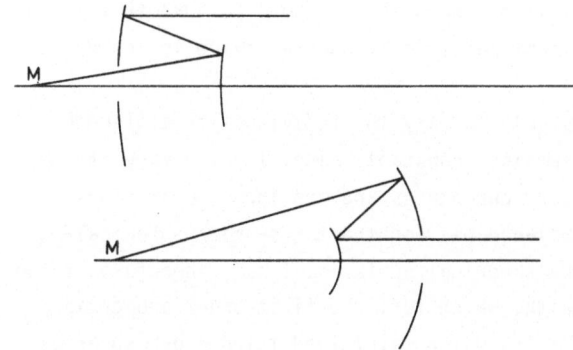

Fig. 1.1. Cassegrain-like
systems. The main focus is
at M

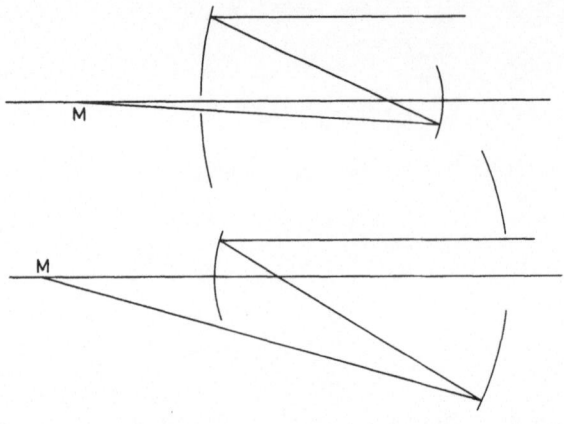

Fig. 1.2. Gregorian-like systems. The main focus is at M

parameter as the *power parameter*. The second parameter, which enters in a highly complicated way, leaves the focal length unchanged but alters the curvatures and the axial separation of the two surfaces, as well as the location of the main focus. This we call the *shape parameter*.

In due course, a paper was prepared by this author [1.1] for publication. Even before it appeared, the possibility of a practical application presented itself. Someone wanted a device that would expand the diameter of a laser beam one hundredfold. Although the specifications called for the system to be diffraction limited, the fact that the laser's radiation was at 10.6 μm made this requirement relatively easy to satisfy. Moreover, the instrument was clearly an on-axis device, so that spherical aberration was the single most important image error to be corrected. Because of the wavelength of the laser's output, a reflecting system was clearly in order. I placed two of my Cassegrain-like systems back to back so that the two main foci coincided and then set the ratio of their two focal lengths at 100. I had then an afocal system with the required magnification and with the added feature that third-order spherical aberration was identically zero. Such a device is shown in Fig.1.3.

But more could be done. I was able to vary the individual focal lengths at will so long as their ratio remained constant. Also, I could vary the two shape parameters, altering the four curvatures and the three separations without upsetting the delicate balance between the third-order spherical-aberration contributions from the spherical surfaces. I had, therefore, three independent parameters to play with, which left the first-order properties of the composite system invariant and which maintained third-order spherical aberration equal to zero. While these three parameters were varied, one at

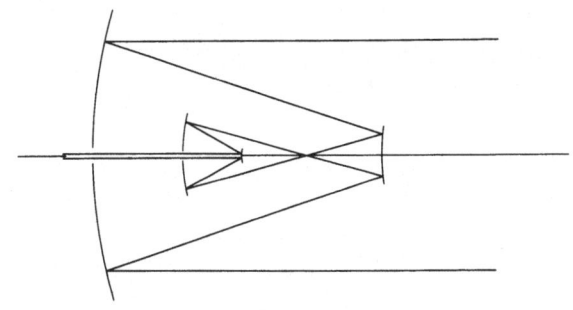

Fig. 1.3. Double Cassegrain afocal system

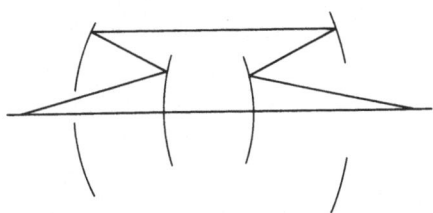

Fig. 1.4. Two joined Cassegrain systems; finite conjugates [1.2]

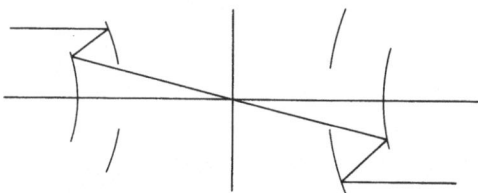

Fig. 1.5. Two joined Cassegrain systems; afocal configuration [1.2]

a time, the behavior of the systems was monitored by ray tracing until it was certain that the diffraction limit was exceeded. Moreover, it was possible to make further adjustments in these parameters to assure that the design was reasonable from the fabricator's point of view.

Unfortunately, the best system was too large for the space available. One of the spherical mirrors was replaced by a plane and another by a paraboloid to obtain a system that would fit the available space. It should be noted that these changes took place in the very last stage of the design process.

Incidentally, MERCADO [1.2] subsequently reinvestigated the problem of the two-mirror system from a more sophisticated point of view and worked out its coupling properties. These are illustrated in Figs.1.4 and 1.5 and will be discussed in greater detail in a later chapter.

The idea of coupling these optical elements together appeared to be a good one. Also, there seemed to be no good reason for these objects to be restricted to reflecting systems. The problem was restated for refracting spherical surfaces. Moreover, Seidel aberrations other than spherical aberration were calculated. In due course, a more general solution was obtained that contained all of the features of the reflecting system but depended on three refractive indices in addition to the shape parameter and the power parameter.

At this point, it became apparent that an even stronger statement could be made. The curvatures and separations of the two surfaces, as well as the location of the main focus, were determined by the same two parameters, as was the reflecting system. Now it became evident that a pair of conjugate pupil planes, determined in the same way, would result in third-order astigmatism vanishing for every choice of chief ray. These results were dutifully reported in a second paper and a more detailed technical report [1.3,4].

The idea of a modular approach to optical design began to take shape. Any number of these two-surface objects, which we will now refer to as *modules*, evidently can be assembled in such a way that adjacent foci and pupil planes coincide, resulting in an optical system for which third-order spherical aberration *and* third-order astigmatism are identically zero. Moreover, a number of power parameters and shape parameters can be adjusted without upsetting these zero values and which would, moreover, not disarray the first-order properties of the system.

Modules then have some of the properties of the thin lens used as a design tool. Modules depend on a power parameter and on a shape parameter. However, the module is thick, not thin, and its power parameter represents true power and not the thin-lens approximation. Like thin lenses, modules can be assembled to form trial layouts. Unlike thin lenses, the modular layouts describe systems in terms of their true power. In that the shape parameter of a module cannot vary the module's power, the shape parameter and the power parameter are truly orthogonal.

However, before modules can be considered as useful tools in optical design, a number of problems need to be solved. Modules are more like elephants than boxcars. Boxcars commute; they can be coupled together in any order. Elephants, on the other hand, if they are to go anywhere, must be coupled trunk to tail. The module's third-order aberrations are calculated with respect to its main focus, whose image is at infinity. Clearly, two finite focal points can be made to coincide, as can two infinitely remote image points. But under no circumstances can a finite focus be made to coincide with an infinite object point. Thus, while a coupled boxcar can be uncoupled, rotated, and recoupled, an elephant and a module cannot.

To clarify this matter, we distinguish two distinct orientations for a module. A module is said to be in its *forward orientation* when the medium of its main focus is to its left and the conjugate at infinity to its right. Here we use the common convention that light propagates generally from left to right. A module in a *backward orientation* has the location of the main focus and the infinite conjugate reversed. Clearly, a module in a forward

5

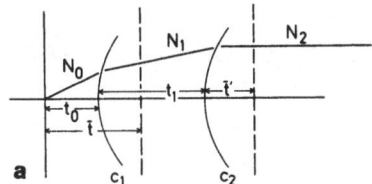

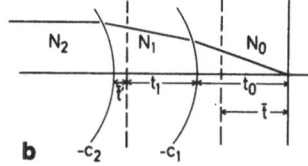

Fig. 1.6. Modules in forward and backward orientation, schematic only. (a) is a module in its forward orientation, (b) in its backward orientation. Dashed lines represent pupil locations

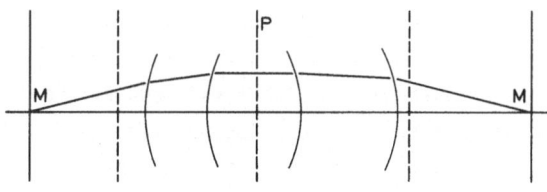

Fig. 1.7. Easy-way coupling; a finite-conjugate system, schematic only. A forward module followed by a backward module. M shows the locations of the two main foci. P is the location of the shared pupil plane

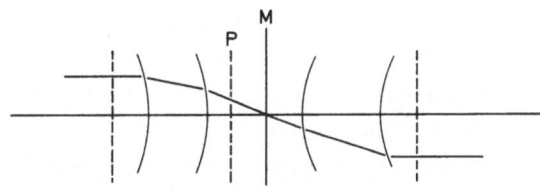

Fig. 1.8. Hard-way coupling; an afocal system, schematic only. A backward module followed by a forward module. M represents the common main focus, P is the shared pupil plane

orientation can be coupled only with a module in a backward orientation. In an optical system constructed out of modules, backward and forward orientations must alternate. Figure 1.6 shows modules in their forward and backward orientations.

The next problem is also concerned with coupling. We have seen that the main foci, or their infinite conjugates, of two adjacent modules must coincide. In addition, to assure that third-order astigmatism for the coupled system is zero, corresponding pupil planes must also coincide. In the case where the two infinite conjugate points are to coincide, the only condition that needs to be satisfied is that at least one of the pupil planes be real; that is to say, it lies outside of the module. We refer to this situation as the *easy-way coupling*; it occurs when a module in a forward orientation is followed by a module in a backward orientation (shown in Fig.1.7). *Hard-way coupling* is more difficult. In this case, a module in a backward orientation is followed by one in a forward orientation. Both the main foci and the appropriate pupil planes must be made to coincide simultaneously. At least one of the main foci and at least one of the two pupil planes must be real. Figure 1.8 shows a hard-way-coupled module.

The coupling problem took several years to solve. The first step toward its solution was taken in 1970 by POWELL [1.5], who used the y-$\bar{y}$ diagram [1.6] to explore ways in which hard-way coupling could be accomplished. It was subsequently shown that there were necessary and sufficient conditions that had to be satisfied by the shape parameters of adjacent modules. Moreover, a hard-way coupling requires the commitment of one of the two power parameters. That is to say, the power parameter of the second module in a hard-way-coupled pair must be proportional to the power parameter of the first module, the factor of proportionality being dependent on the two shape parameters.

The necessary conditions led to the study of the morphological changes of the module that occur as the shape parameter varies. *Critical values* of the shape parameters, values at which the module undergoes a profound change, were found and studied. Most of these turned out to be functions of the three refractive indices. This led to a classification scheme for the module, which describes the coupling relations between the critical values of the shape parameters. In terms of this classification scheme, some of the necessary conditions for coupling are stated. An adequate description of these rather complicated results must be deferred to a later chapter.

Once the problem of coupling was solved, a convenient method of assembling modules into lens systems was needed. Because the power parameter was, after all, only a scale factor, whose value was to be determined by solving some equations, it seemed appropriate to separate it from the other formulas that describe the module. This led to the idea of *canonical modules* and *canonical optical parameters*. These latter are expressions for the curvatures, the separations, and the location of the main focus and the pupil planes in which the value of the power parameter is set equal to unity. These canonical quantities are represented in what follows by capital letters, whereas the real optical parameters and the power parameters are represented by lower-case letters. Thus, if t_j represents the distance between the jth and the $j+1$ surface and f represents the power parameter, then $t_j = fT_j$, where T_j is the appropriate canonical optical parameter. In like manner, $c_1 = C_1/f$, where C_1 is the canonical curvature of the first surface of a module and c_1 is its real counterpart.

From the idea of the canonical module it is a short step to *canonical ray tracing* and *canonical rays*. A *canonical marginal ray* is a marginal ray with a standard slope traced through a forward-oriented canonical module. A *canonical chief ray* is defined in a similar way. The next step is to define *canonical aberration coefficients*, which can then be related in an obvious way to the real aberration coefficients. The relationship between the two usually involves the power parameter raised to some power as a factor. For example, real third-order coma is $1/f^2$ times canonical coma; Petzval and primary axial chromatic aberration are linear in $1/f$, whereas distortion and primary lateral chromatic aberration are independent of f.

Thus, with a system of modules coupled to form an optical system, we can obtain from the canonical aberration equations a set of simultaneous equations in the power parameters and shape parameters of the individual modules. The expression for zero coma, for example, turns out to be a quadratic in $1/f^2$; for zero Petzval and primary axial color each are linear, whereas those for distortion and lateral color lead to equations in the shape parameters alone [1.7,8].

This arrangement suggests a reasonable *modus operandi* for assembling modules into an optical design. The two equations that govern distortion and lateral color can first be solved for appropriate values of the shape parameters to yield a system for which these aberrations are zero but for which the power parameters are yet undetermined. Then the remaining two equations can be solved for the power parameters that yield a system for which all Seidel aberrations and the primary chromatic aberrations are zero. However, things do not always work out so easily and, at the appropriate place, certain necessary and sufficient conditions that assure the existence of solutions will be discussed. ANDERSON [1.9,10] has extended these results to the fifth-order aberrations, which are far more complicated but, nevertheless, yield equations that, in principle, can be solved.

The equations for modular optical design, like any formal procedure for optical design, are complicated. Yet the concept is rather simple. (Some would say simple minded.) We have found that the best approach to the process is with the use of an interactive computer, such as a time-sharing computer, which can be interrupted while running and parameters altered or modules added. Indeed, almost all of the experimental work done in this research made use of the General Electric time-sharing service and the University of Arizona DEC-10 computer. Because of its simplicity and elegance, as well as the ease with which a novice can learn to use it, the BASIC computer language was used exclusively. Moreover, it is necessary to look to the future. Inexpensive microcomputers are now not only available but abundantly so. Timesharing computers, except in certain circumstances are obsolescent.

1.2 Paraxial Optics

The theory of optical-design modules depends entirely on paraxial optics and the Seidel aberrations. Although I would anticipate that anyone reading this volume would be well versed on these subjects, in order to fix ideas and settle on a convenient notation, they will be summarized. We will follow, more or less, WELFORD's book [1.11].

Paraxial ray tracing is an approximation to real or finite ray tracing. One way of looking at the subject is to start with the formulas for real ray tracing, replace square roots and trigonometric functions by their power-series representations and then truncate all but the linear terms. The region of validity of the approximation is, strictly speaking, limited to an infinitesimal tube that contains the axis of the rotationally symmetric optical system. However, in practice, paraxial rays are by no means confined to this region.

Where direction cosines or reduced direction cosines are used to determine the orientation of a real ray in space, the direction of its paraxial counterpart is determined by its slope. The intersection of a paraxial ray and a refracting surface is represented as taking place on a plane tangent to the surface and perpendicular to the axis of symmetry of the lenses.

Real rays divide themselves neatly into two categories. Meridian rays are rays confined to the meridian plane of the optical system. This plane is the unique plane that contains the object point and the axis of symmetry. Skew rays are rays that are not meridian rays. SMITH [1.12] defined a quantity, the skewness invariant, as a measure of the amount a skew ray fails to be a meridian ray (see also [1.13]). Suffice it to say that the skewness invariant is a quadratic form in the coordinates and direction cosines of a real ray, so that its paraxial approximation is therefore zero. We may conclude that all paraxial rays are necessarily meridian rays.

In what follows, lower-case letters are used to denote the elements of a lens and the parameters associated with a paraxial ray in that lens. Subscripts are used to indicate the location of elements and rays. Thus, c_j represents the curvature of the jth surface, and y_j the height of a paraxial ray on that surface; N_j and D_j are the refractive index and dispersion of the medium that follows that surface, respectively; and u_j is the slope of a ray in that medium. The geometric distance between the jth surface and the next-following surface is given by t_j.

The paraxial-ray-tracing formula for refraction is, following WELFORD [Ref.1.11, Sect.3.10],

$$N_{j+1}u_{j+1} = N_j u_j - (N_{j+1}-N_j)c_{j+1}y_{j+1} \quad . \tag{1.2.1}$$

In a more compact form, we may write this as

$$u_{j+1} = \mu_{j+1}u_j - (1-\mu_{j+1})c_{j+1}y_{j+1} \quad , \tag{1.2.2}$$

where

$$\mu_{j+1} = N_j/N_{j+1} \quad . \tag{1.2.3}$$

Note that the terms of (1.2.1) can be rearranged so that

$$N_j(c_{j+1}y_{j+1}+u_j) = N_{j+1}(c_{j+1}y_{j+1}+u_{j+1}) \quad .$$

We can see that $c_{j+1}y_{j+1}+u_j$ is the paraxial angle of incidence, that $c_{j+1}y_{j+1}+u_{j+1}$ is the paraxial angle of refraction, and that the above equation is the paraxial version of Snell's law. We define, again following WELFORD [Ref.1.11, Sect.5.2], the paraxial refraction invariant:

$$a_{j+1} = N_j(c_{j+1}y_{j+1}+u_j) = N_{j+1}(c_{j+1}y_{j+1}+u_{j+1}) \quad . \tag{1.2.4}$$

The paraxial-ray-tracing equation for the transfer operation is

$$y_{j+1} = y_j + t_j u_j \quad . \tag{1.2.5}$$

Next, we cast the equations for transfer and refraction, (1.2.5) and (1.2.2), respectively, in matrix form, in the manner of BROUWER [1.14]:

$$\begin{pmatrix} y_{j+1} \\ u_j \end{pmatrix} = \begin{pmatrix} 1 & t_j \\ 0 & 1 \end{pmatrix} \begin{pmatrix} y_j \\ u_j \end{pmatrix} \quad , \tag{1.2.6}$$

$$\begin{pmatrix} y_{j+1} \\ u_{j+1} \end{pmatrix} = \begin{pmatrix} 1 & 0 \\ -c_{j+1}(1-\mu_{j+1}) & \mu_{j+1} \end{pmatrix} \begin{pmatrix} y_{j+1} \\ u_j \end{pmatrix} \quad . \tag{1.2.7}$$

These we combine, to obtain a matrix product that represents a transfer followed by a refraction,

$$\begin{pmatrix} y_{j+1} \\ u_{j+1} \end{pmatrix} = \begin{pmatrix} 1 & t_j \\ -c_{j+1}(1-\mu_{j+1}) & \mu_{j+1}-t_j c_{j+1}(1-\mu_{j+1}) \end{pmatrix} \begin{pmatrix} y_j \\ u_j \end{pmatrix} \quad . \tag{1.2.8}$$

From these calculations, we can see that, in any medium, the aggregate of all paraxial rays constitutes a two-dimensional vector space. Any three paraxial rays can be seen to be linearly dependent. It is therefore possible to select two basis elements from which, by forming all possible linear combinations, we can generate all other paraxial rays. The two rays that are chosen to form the basis are the marginal and chief paraxial rays. In the notation that we will use, the chief ray and all quantities associated with it will be indicated by placing a bar over the appropriate symbols. Thus $\bar{y}$ and $\bar{u}$ denote, respectively, the height and the slope of a paraxial chief ray; y and u denote the same quantities for the paraxial marginal ray. The marginal ray is the ray from the axial object point that just clears all obstructions in the optical system, such as stops and edges of lens elements. The chief ray is the ray from the edge of the object field that passes through the center of the stop.

An important datum in the calculations that follow is the Lagrange invariant [1.11],

$$h = N_j(\bar{u}_j y_j - u_j \bar{y}_j) \quad . \tag{1.2.9}$$

By substituting the formulas for refraction and transfer, (1.2.2) and (1.2.5), into (1.2.9), we can see that h is invariant with respect to refraction and transfer. Once the paraxial marginal and paraxial chief rays are chosen, h is constant at every surface and in every medium in the lens. Moreover, referring now to the definition of the refraction invariant, (1.2.4), we can see that

$$h = \bar{a}_j y_j - a_j \bar{y}_j \quad . \tag{1.2.10}$$

WELFORD [1.11] uses the symbol Δ to denote the difference of a quantity calculated on either side of a refracting surface. Thus we can indicate that the Lagrange invariant h and the refraction invariant a are indeed invariant by the expressions $\Delta h = 0$ and $\Delta a = 0$, respectively.

Other quantities that we will use are

$$\Delta(u_j/N_j) = u_{j+1}/N_{j+1} - u_j/N_j \quad .$$

It is most convenient to denote these quantities by s_{j+1}, so that using (1.2.2) we obtain

$$s_{j+1} = -(1-\mu_{j+1})[(1+\mu_{j+1})u_j + \mu_{j+1}c_{j+1}y_{j+1}]/N_j \quad . \tag{1.2.11}$$

We next calculate the Seidel aberration coefficients. Again we refer to [Ref.1.11, Sect.6.7]; however, we will use a slightly different notation for the Petzval contribution. The image errors come first.

Spherical aberration: $\quad b_j = -a_j^2 y_j \Delta(u_{j-1}/N_{j-1}) = -a_j^2 y_j s_j \quad ;$ $\qquad$ (1.2.12)

Coma: $\qquad\qquad\qquad \mathcal{b}_j = -a_j \bar{a}_j y_j s_j \quad ;$ $\qquad\qquad$ (1.2.13)

Astigmatism: $\qquad\qquad c_j = -\bar{a}_j^2 y_j s_j \quad ;$ $\qquad\qquad$ (1.2.14)

Petzval: $\qquad\qquad\quad p_j = -c_j \Delta(1/N_{j-1}) = c_j(1-\mu_j)/N_j \quad ;$ $\qquad$ (1.2.15)

Distortion: $\qquad\qquad e_j = -[\bar{a}_j^2 y_j \Delta(u_j/N_j) + h^2 c_j \Delta(1/N_j)]\bar{a}_j/a_j$

$$= \bar{a}_j(c_j + h^2 p_j)/a_j \quad . \tag{1.2.16}$$

WELFORD [Ref.1.11, Sect.5.3] defines the eccentricity parameter e by

$$\bar{y} = yhe \quad , \tag{1.2.17}$$

so that

$$e = \bar{y}/hy \quad . \tag{1.2.18}$$

Now, referring to the refraction invariant defined in (1.2.4), we may obtain

$$\bar{a} = h(1+aye)/y \quad .$$

Then the pupil aberrations can be written in the form:

Spherical aberration: $\bar{b}_j = -e_j(he_j + \bar{a}_j\bar{y}_j p_j)$; $\qquad$ (1.2.20)

Coma: $\qquad\qquad \bar{\delta}_j = e_j + h\Delta(\bar{u}_j^2)$; $\qquad$ (1.2.21)

Astigmatism: $\qquad \bar{c}_j = c_j + h\Delta(u_j u_j)$; $\qquad$ (1.2.22)

Distortion: $\qquad \bar{e}_j = \bar{\delta}_j + h\Delta(u_j^2)$. $\qquad$ (1.2.23)

The Petzval contribution is not included in this list. Because it is a first-order quantity and therefore independent of the paraxial rays, the Petzval contribution is the same for image and pupil.

The quantities $\bar{b}_j$, $\bar{\delta}_j$, $\bar{c}_j$, and $\bar{e}_j$ can be written in yet another form. If we use the formulas for b_j, δ_j, c_j, and e_j, found in (1.1.12-16), and inter-change u and $\bar{u}$, and y and $\bar{y}$, we obtain formulas for the barred quantities. In doing this, we must be careful to take into account that the Lagrange invariant changes sign under this transformation.

Now we may write the pupil aberrations in the form

$$\bar{b}_j = -\bar{a}_j^2\bar{y}_j\bar{s}_j \ , \qquad\qquad (1.2.24)$$

$$\bar{\delta}_j = -a_j\bar{a}_j\bar{y}_j\bar{s}_j \ , \qquad\qquad (1.2.25)$$

$$\bar{c}_j = -a_j^2\bar{y}_j\bar{s}_j \ , \qquad\qquad (1.2.26)$$

$$\bar{e}_j = a_j(\bar{c}_j - hp_j)/\bar{a}_j \ . \qquad\qquad (1.2.27)$$

In our calculations, we will prefer these formulas to those given in (1.2.20-23); that is, we prefer these, with a pair of exceptions. We replace the expression in (1.2.6) by (1.2.21),

$$e_j = \bar{\delta}_j - h\Delta(\bar{u}_j^2) \ , \qquad\qquad (1.2.28)$$

where $\bar{\delta}_j$ is given in (1.2.25). We will use this expression for image distortion. We do exactly the same thing for pupil distortion, using (1.2.23). Here we use (1.2.13) to calculate δ_j.

1.3 Primary Chromatic Aberrations

Somewhere along the line, geometrical optics must come to grips with at least one aspect of the physics of refraction. Chromatic aberrations arise from the fact that the refractive indices of most optical materials depend on the wavelength of light. Thus, a ray of white light (to use extremely loose terminology, to say the least) would decompose at each refracting surface into a spectrum of monochromatic rays, each of which heads in a slightly different direction and at a slightly different speed determined by its own wavelength. This effect, compounded at each refracting surface in the optical system, will, unless certain precautions are taken, result in an unsightly mess on the image plane.

This description, as far as physical accuracy is concerned, is at best whimsical. Nevertheless, in it lies the bare bones of the approach to the problem of chromatic aberration. In the design of optics for use over the visible wavelength range, the designer selects some central wavelength. At this wavelength, the refractive indices of the glasses to be used in the design are found. This kind of information is invariably provided by manufacturers of optical glass. For example, Schott normally provides central-wavelength data at the d wavelength, the yellow helium line ($\lambda=0.58756$ μm), and at the e wavelength, the green mercury line ($\lambda=0.54607$ μm). Because of its brightness, the D line, the center of the yellow sodium doublet ($\lambda=0.58929$ μm), is frequently used. Refractive indices at these lines are denoted by the symbols N_d, N_e, and N_D, respectively.

A value of *dispersion* is normally also required. Dispersion is the difference between two refractive indices at wavelengths greater than and less than the central wavelength. This range of wavelengths is the region over which the chromatic aberrations of the lens are to be corrected. For use with the d central wavelength, Schott provides the dispersion $N_F - N_C$, where the F wavelength is the blue hydrogen line ($\lambda=0.48613$ μm) and the C wavelength is the red hydrogen line ($\lambda=0.65627$ μm). For the e central wavelength, they provide the dispersion $N_{F'} - N_{C'}$, where F' represents the blue cadmium line ($\lambda=0.47999$ μm) and where C' corresponds to the red cadmium line ($\lambda=0.64385$ μm). With the data provided, we can construct any combination of indices to cover any combination of preferred wavelengths. Although additional data are frequently provided, for our purposes we will need the refractive index at a central wavelength, N, and the dispersion at the wavelength range, D.

The two primary chromatic aberrations are the only chromatic aberrations we will consider. Again, we draw on WELFORD [Ref.1.11, Sect.9.5]:

Longitudinal chromatic aberration: $\ell = ayk$,   (1.3.1)

Transverse chromatic aberration: $\ell = \bar{a}yk$.   (1.3.2)

Here k is given by

$$k = \Delta(D_j/N_j) = D_{j+1}/N_{j+1} - D_j/N_j \quad . \tag{1.3.3}$$

The first of these, longitudinal chromatic aberration, is essentially a change of focal length with wavelength. Transverse chromatic aberration or lateral chromatic aberration can be regarded as a change of magnification.

With these two chromatic aberrations, we conclude the catalog of aberration formulas that we will use in the immediately succeeding chapters. Much later, we will come to grips with the fifth-order aberrations.

2. The y-ȳ Diagram

2.1 Definitions and General Description

One of the most intriguing developments in the field of the geometrical optics of rotationally symmetric optical systems is the Delano y-ȳ diagram [2.1],
see also [2.2,3]. Picture a lens that consists of several refracting surfaces
that separate media with prescribed refractive indices. We choose a coordinate
system in which the z axis is the axis of symmetry. Let the y,z plane be the
meridian plane, which, in Fig.2.1, is the plane of the paper. We will be concerned with paraxial rays exclusively and will have no need of an x axis.

We begin at some axial object point and trace a marginal paraxial ray
through the system. It is refracted at each surface, grazes the edge of the
stop, emerges in image space and proceeds with no further ado to the image
plane, where it intersects the axis. The ray path consists of a sequence of
connected straight lines, the intersections of which are associated with the
refracting surfaces. However, we must remind ourselves that this is a paraxial ray and the intersections, speaking strictly, are not at the refracting

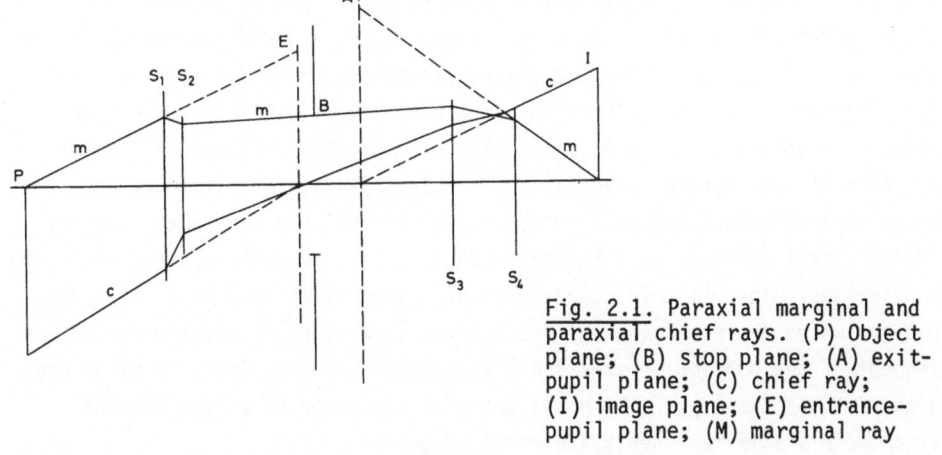

Fig. 2.1. Paraxial marginal and
paraxial chief rays. (P) Object
plane; (B) stop plane; (A) exit-
pupil plane; (C) chief ray;
(I) image plane; (E) entrance-
pupil plane; (M) marginal ray

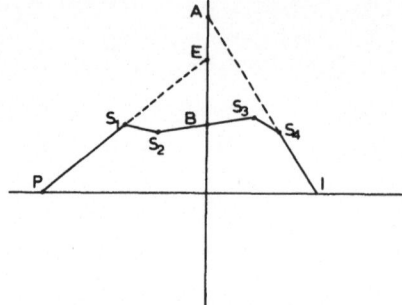

Fig. 2.2. y-ȳ diagram, representing the
lens shown in Fig.2.1. (P) Object plane;
(B) stop plane; (A) exit-pupil plane;
(I) image plane; (E) entrance-pupil plane

surfaces but lie on a plane tangent to the refracting surfaces at their axial
points. Note that, wherever a line segment crosses the axis, an image is
formed. When an extension of a line segment intersects the axis, the image
formed there is virtual.

We represent the ray path as a functional relationship, by writing its
height from the axis, y, as a function of its distance along the axis. Thus,
y = y(z) describes a broken straight line on the y,z plane.

Exactly the same thing is done with the chief ray. It begins at the edge
of the field, is refracted at each surface, passes through the center of the
stop, emerges into image space and arrives at the image plane. Again, we have
a broken line with corners that represent the refracting surfaces, which we
write functionally as ȳ = ȳ(z). Wherever a segment crosses the axis, we have
there an image of the stop. If an extension of the line segment crosses the
axis, then the image there of the stop is virtual. In object space or image
space, the paraxial image of the stop is the entrance pupil or the exit pupil.

To form the y-ȳ diagram for this lens, we simply plot y versus ȳ, treating z as a free parameter. We obtain another curve that consists of a sequence
of connected straight-line segments. Figure 2.2 is the y-ȳ diagram of the lens
shown in Fig.2.1. Again, the corners correspond to refracting surfaces. We
find that the curve tends to circulate around the origin to form an open
polygon. Whether the circulation is clockwise or counterclockwise depends on
the sign of the Lagrange invariant. The object plane is represented by a
point that lies on the ȳ axis and marks the beginning of the open polygon.
If the object point is at infinity, then the first segment is parallel to the
ȳ axis. The image plane is represented by the point at the end of the open
polygon where the final segment crosses the ȳ axis. Every point where a line
segment of the polygon crosses the ȳ axis marks the location of a plane conjugate of the image plane. A point where an extension of a line segment
crosses this axis represents a virtual image.

Pupil planes and stop planes are treated in exactly the same way. The entrance pupil is represented by the point where the initial line segment of the polygon, or its extension, intersects the y axis. The point where a segment cuts the y axis locates the stop or a plane conjugate to the stop. Where the final segment or its extension cuts the y axis corresponds to the exit pupil.

Recall from (1.2.9) the Lagrange invariant, $N(\bar{u}y - u\bar{y}) = h$, a constant at every surface and in every medium of the lens. By writing it in the form

$$y(N\bar{u}/h) + \bar{y}(-Nu/h) = 1 \quad , \tag{2.1.1}$$

we have it in the intercept form of the equation of a straight line. This is the equation of a line segment of the y-$\bar{y}$ diagram between two corners. Its $\bar{y}$ and y intercepts are given, respectively, by

$$\bar{y} = h/Nu \quad , \quad y = -h/N\bar{u} \quad . \tag{2.1.2}$$

The slope of the line segment is therefore

$$m = u/\bar{u} \quad . \tag{2.1.3}$$

From this we can see that from the y-$\bar{y}$ diagram alone, provided that the Lagrange invariant is known, we can obtain the values of the marginal and chief rays directly.

2.2 Conjugate Lines

Next consider a straight line through the origin of the diagram. We observe that the ratio y/$\bar{y}$ is constant along the line; indeed, it is exactly equal to its slope.

Now consider the optical system represented by the y-$\bar{y}$ diagram. Choose any pair of conjugate planes in the system. Let the magnification relationship between these planes be given by the constant M such that is y is the height of a marginal ray on the first plane then My will be its height on the second. The same is true of the chief ray. If $\bar{y}$ is its height on the first plane, then M$\bar{y}$ must be its height on the second. It follows that the ratio y/$\bar{y}$ must be constant on every pair of conjugate planes.

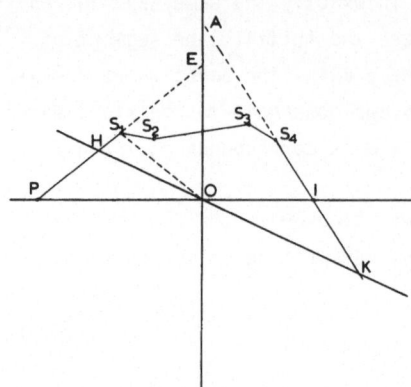

Fig. 2.3. Conjugate points; conjugate
lines. H and K represent conjugate planes
in object and image space, respectively.
The area of triangle HOS is proportional
to the distance from H to the first surface. HOK is called the conjugate line

Next we look at each component of the optical system. The object plane is
mapped by the first refracting surface onto some conjugate plane on which
the ratio $y/\bar{y}$ will have the same value as it did on the object plane. The
second surface defines a plane conjugate also to the object plane on which
the ratio assumes the same value. And so its goes through the optical system,
until the image plane is reached. The point of all this is that the ratio $y/\bar{y}$
is constant on all sets of conjugate planes. It follows that, in a $y-\bar{y}$ diagram, conjugate points must lie on a straight line through the origin. This
is shown in Fig.2.3.

We call such a straight line a *conjugate* line. Clearly, the two coordinate
axes are special cases of conjugate lines, one being associated with the object and image plane and the other with the stop plane and the pupil planes.

2.3 Axial Distances

Finally, we consider the triangle formed by two corners of the $y-\bar{y}$ diagram
polygon and the origin. Two of the sides of the triangle are conjugate lines
whereas the third consists of a line segment of the polygon. Let the coordinates of the corners be $(\bar{y}_1, y_1)$ and $(\bar{y}_2, y_2)$. Then, by a well-known formula,
the area of the triangle is given by

$$A = (\bar{y}_2 y_1 - y_2 \bar{y}_1)/2 \quad .$$

(2.3.1)

The two corners represent two locations of planes in the optical system. Let the axial distance between the two planes be z so that, from the transfer equation, (1.2.5), we get

$$y_2 = y_1 + zu \quad , \quad \bar{y}_2 = \bar{y}_1 + z\bar{u} \quad ,$$

where u and $\bar{u}$ are the slopes of the marginal and chief rays between these two planes. Applying these to the expression for area, we obtain

$$A = z(\bar{u}y_1 - u\bar{y}_1)/2 \quad ,$$

which leads at once to

$$z = 2NA/h \quad , \tag{2.3.2}$$

where h is the Lagrange invariant given in (1.2.9).

We have shown several things here. We may read directly the heights of the marginal and chief rays, y and $\bar{y}$, at each refracting surface from the coordinates of the corners of the open polygon in the y-$\bar{y}$ diagram. The intersection of the line segment that joins a pair of corners, or its extension, determines the slope of the marginal and chief ray between these two surfaces. The axial distances between two refracting surfaces can be determined from the area of the triangle subtended by the origin and two consecutive corners of the polygon, as shown in Fig.2.4.

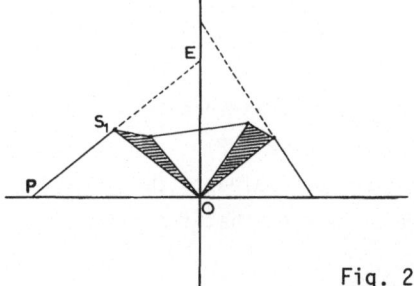

Fig. 2.4. Triangle areas and axial distances. Areas of cross-hatched triangles are proportional to element thicknesses. Area of triangle POS$_1$ is proportional to object distance. Area of triangle POE represents distance to virtual entrance-pupil plane

2.4 Power

Next, consider a corner point $(\bar{y},y)$ on a y-$\bar{y}$ diagram. We know that this represents a refracting surface. Recall that the refraction formula from (1.2.1) is

$$N_2 u_2 = N_1 u_1 - (N_2 - N_1) c y \quad .$$

We speak of the power of a refracting spherical surface as

$$\Phi = (N_2 - N_1) c \quad , \tag{2.4.1}$$

where c represents the curvature of the sphere.

Here we should remember that Φ is a thin-lens approximation of the real power of the surface. Using (2.4.1), we may write the equations for the refraction of a marginal ray and a chief ray in the form

$$N_2 u_2 = N_1 u_1 - \Phi y \quad , \quad N_2 \bar{u}_2 = N_1 \bar{u}_1 - \Phi \bar{y} \quad .$$

Eliminating y and $\bar{y}$ between these two equations and applying the Lagrange invariant, (1.2.10), we get

$$\Phi = N_2 N_1 (u_1 \bar{u}_2 - \bar{u}_1 u_2)/h \quad . \tag{2.4.2}$$

This may also be written in the form

$$\Phi = h \begin{vmatrix} N_1 u_1/h & N_2 u_2/h \\ N_1 \bar{u}_1/h & N_2 \bar{u}_2/h \end{vmatrix} \quad , \tag{2.4.3}$$

which involves a determinant that involves the y and $\bar{y}$ intercepts of the two line segments that meet at $(\bar{y},y)$. These are found in (2.1.2).

The power of this refracting surface can also be related to the angle between the two line segments (see Fig.2.5). We have seen that the slopes of the two line segments are, from (2.1.3),

$$m_1 = u_1/\bar{u}_1 \quad \text{and} \quad m_2 = u_2/\bar{u}_2 \quad .$$

It follows that the angle θ between these two lines is given by

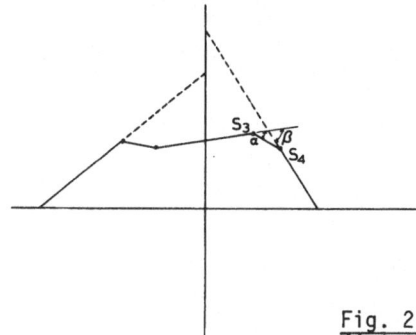

Fig. 2.5. Angles and powers. Angle α is used with (2.4.4) to calculate the power of surface S_3. Angle β is used to calculate the power of the component that consists of surfaces S_3 and S_4

$$\tan\theta = \frac{m_2 - m_1}{1 + m_2 m_1} = \frac{u_2\bar{u}_1 - \bar{u}_2 u_1}{\bar{u}_2\bar{u}_1 + u_2 u_1} \quad .$$

Applying to this (2.4.2), we obtain

$$\tan\theta = h\phi/N_2 N_1(\bar{u}_2\bar{u}_1 + u_2 u_1)$$

or

$$N_2 N_1(\bar{u}_2\bar{u}_1 + u_2 u_1) = h\phi \cot\theta \quad .$$

Now consider this equation and (2.4.2) as a simultaneous pair and solve for u_2 and $\bar{u}_2$, obtaining

$$u_2 = h\phi(u_1 \cot\theta + \bar{u}_1)/N_2 N_1(u_1^2 + \bar{u}_1^2) \quad ,$$

$$\bar{u}_2 = h\phi(\bar{u}_1 \cot\theta - u_1)/N_2 N_1(u_1^2 + \bar{u}_1^2) \quad .$$

The sum of the squares of these two equations leads at once to

$$N_2^2 N_1^2(u_2^2 + \bar{u}_2^2)(u_1^2 + \bar{u}_1^2) = h^2\phi^2 \csc^2\theta \quad .$$

We may therefore write the power of the refracting surface as

$$\phi = \sin\theta \sqrt{\left(\frac{N_2 u_2}{h}\right)^2 + \left(\frac{N_2\bar{u}_2}{h}\right)^2} \sqrt{\left(\frac{N_1 u_1}{h}\right)^2 + \left(\frac{N_1\bar{u}_1}{h}\right)^2} \quad . \tag{2.4.4}$$

Note that the squared quantities under the radical are, by (2.1.2), re-
ciprocals of the intercepts on the two line segments that join at the corner
point $(\bar{y}, y)$.

This suggests a reciprocal relationship, which is seen most clearly in
(2.1.1), the intercept form of the equation for the line segment in a $y-\bar{y}$
diagram. This had led LOPEZ-LOPEZ [2.4] to the study of what he calls $\omega-\omega$
diagrams, where $\omega = Nu$.

2.5 Stop Shifts. Conjugate Shifts

The orthogonality of the coordinate axes of the $y-\bar{y}$ diagram is convenient
only because we are used to orthogonal-coordinate systems. The two axes are
two conjugate lines that just happen to be the conjugate lines for the object and the stop. The intersection of any line segment of the $y-\bar{y}$ diagram,
or its extension, with the $\bar{y}$ axis represents a plane conjugate to the object
plane. In like manner, the intersection of such a line with the y axis represents a plane conjugate to the stop plane. It is clear that all pupils
are represented by points on this line.

Suppose we choose to move the stop a distance z in, say, the triplet whose
$y-\bar{y}$ diagram is shown in Fig.2.6. To illustrate this change in the $y-\bar{y}$ diagram,
we need to find the triangle whose vertex is at the origin, whose side coin-
cides with the $\bar{y}$ axis, and whose area is proportional to the displacement z.
The exact area required is given by (2.3.2),

$$A = hz/2N \quad , \qquad\qquad\qquad (2.5.1)$$

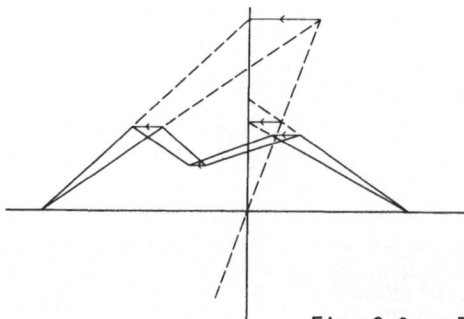

Fig. 2.6. $y-\bar{y}$ diagram of lens showing stop shift

where h is the Lagrange invariant and where N is the refractive index of the medium in which the stop is located. This is presumably air, so $N = 1$ for the overwhelming majority of cases. Having found the area A, we next use (2.3.1) to determine one of the coordinates of the intersection of the conjugate line and the appropriate line segment. Because one point of the triangle lies on the y axis, $\bar{y}_1 = 0$ and

$$\bar{y}_2 = 2A/y_1 \quad .$$

For the other coordinate, we use the intercept form of the equation of the line segment, (2.1.1),

$$\bar{y}(Nu) - y(N\bar{u}) = h \quad ,$$

in which we substitute the formula for $\bar{y}_2$. We end up with the following expressions for $\bar{y}_2$ and y_2:

$$\bar{y}_2 = 2A/y_1$$
$$y_2 = 2A/N\bar{u}y_1 - h/N\bar{u} \quad . \tag{2.5.2}$$

These two coordinates provide the point on the line segment through which the conjugate line must pass. This is the dashed line through the origin in Fig.2.6. The pupil positions that correspond to the new location of the stop can be calculated from the y-$\bar{y}$ diagram and (2.3.2).

To restore the y-$\bar{y}$ diagram to its pristine form, in which the axes are orthogonal, we need only apply a shear transformation parallel to the $\bar{y}$ axis to each point of the diagram. This is shown in Fig.2.6.

A conjugate shift is done in exactly the same manner. Here, a conjugate line is found which forms a triangle with the $\bar{y}$ axis and the appropriate line

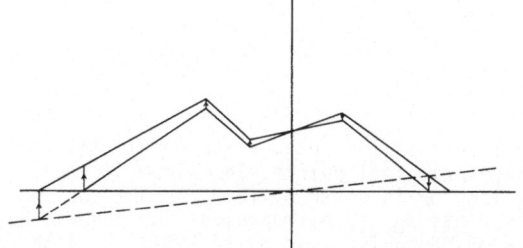

Fig. 2.7. y-$\bar{y}$ diagram of lens showing conjugate shift

segment, with an area proportional to the amount of the shift. Again, (2.3.2) is invoked. Equation (2.3.1) is then used to find one of the coordinates of the point of intersection. The Lagrange invariant in the slope-intercept form of the equation of the straight line is used to determine the second coordinate. Normally, this step is followed by a shear transformation parallel to the y axis to restore the y-$\bar{y}$ diagram to its familiar orthogonal arrangement. Figure 2.7 shows how this is done.

2.6 Cardinal Points

The cardinal points of an optical system — the two foci, the principal points, and the nodal points — are best illustrated by means of the y-$\bar{y}$ diagram and the idea of the conjugate line. The cardinal points of a lens comprise its overall properties and may be used to represent the lens, without necessarily referring to its constituent elements. Indeed, the cardinal points epitomize the first-order properties of a lens and permit us to take a black-box approach to its description. In what follows, we will adhere to this approach and in Fig.2.8 show only the object line and image line of the y-$\bar{y}$ diagram representation of a lens.

We begin with the principal planes. These are defined as the unique pair of conjugate planes, one in object space and one in image space, for which the magnification is exactly unity. The principal points are the axial points

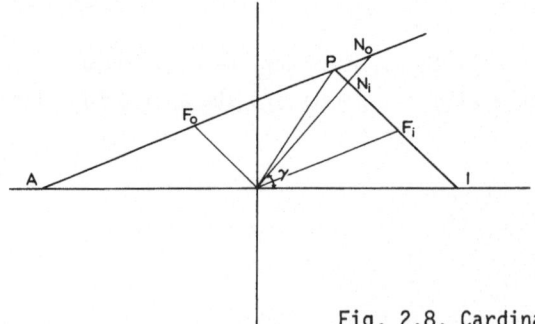

Fig. 2.8. Cardinal points. P represents the object and image principal points. The slope of the conjugate line OP is given by (2.6.2). The two conjugate points N_0 and N_i represent the object- and image-space nodal points. The slope of ON_IN_0 is given by (2.6.3). The line F_0O is parallel to PI. F_0 represents the object-space focus. OF_I is parallel to AP. F_I represents the image-space focus

of the principal planes. Let y_p represent the height of a marginal ray on the object-space principal plane and let y_p' represent its counterpart on the image-space principal plane. Then, because the magnification is unity, it must be that $y_p' = y_p$. A similar equation holds for the chief ray, so that $\bar{y}_p' = \bar{y}_p$. It follows that the two points on the y-$\bar{y}$ diagram, $(\bar{y}_p, y_p)$ and $(\bar{y}_p', y_p')$, must coincide. Where they coincide is the point where the object line and the image line intersect, as shown in Fig.2.8.

The location of the principal planes relative to object plane, image plane, either pupil plane, or any other known reference point in the lens displayed on the y-$\bar{y}$ diagram can be determined by measuring the coordinates of the principal point and using (2.3.1-2).

Making use of the intercept form of the object line and the image line, as in (2.1.1), we obtain

$$y = \bar{y}u/\bar{u} - h/N\bar{u} \quad , \quad y' = \bar{y}'u'/\bar{u}' - h/N'\bar{u}' \quad , \tag{2.6.1}$$

where h is the Lagrange invariant. We find the coordinates of the point of intersection to be

$$y = h(N'u'-Nu)/NN'(u\bar{u}'-u'\bar{u}) \quad ,$$

$$\bar{y} = h(N'\bar{u}'-N\bar{u})/NN'(u\bar{u}'-u'\bar{u}) \quad ,$$

so that the slope of the conjugate line to the principal-point representation on the y-$\bar{y}$ diagram is

$$H_p = (N'u'-Nu)/(N'\bar{u}'-N\bar{u}) \quad . \tag{2.6.2}$$

The nodal points of a lens are defined as the pair of conjugate axial points that have the following property. Any ray through one nodal point that also passes through the lens must also pass through the other nodal point, with its original direction. In other words, the ray that enters the lens through one nodal point emerges from the lens through the second nodal point. Moreover, the emerging ray is parallel to the entering ray. If the refractive indices of object and image spaces are equal, then the nodal points and the principal points coincide. If those refractive indices are not equal, the planes through the nodal points determine conjugate planes with magnification $M = N/N'$, where N and N' are the refractive indices of object space and image space, respectively.

Again, we use (2.6.1), the equations of the object and image lines. Now let the equation of the conjugate line be

$$y = H\bar{y} \quad ,$$

where H is to be determined. The point of intersection of the conjugate line with the object line is $(\bar{y}, y)$, where

$$\bar{y} = -h/N(H\bar{u}-u) \quad , \quad y = -Hh/N(H\bar{u}-u) \quad ,$$

and with the image line, $(\bar{y}', y')$, where

$$\bar{y}' = -h/N'(H\bar{u}'-u') \quad , \quad y' = -Hh/N'(H\bar{u}'-u') \quad .$$

However, we require the magnification to be N/N', so that

$$N'y' = Ny \quad , \quad N'\bar{y}' = N\bar{y} \quad ,$$

from which we determine the slope H_n of the conjugate line in the y-$\bar{y}$ diagram representation of the nodal points to be

$$H_n = (u'-u)/(\bar{u}'-\bar{u}) \quad , \tag{2.6.3}$$

as shown in Fig.2.8.

The remaining two cardinal points are the foci, which are the easiest to see. Each focus is the image of the axial point at infinity. It is necessary only to construct the conjugate line parallel to the object line and to determine its intersection with the image line. This point is then the y-$\bar{y}$ diagram representation of the focus on the image side of the lens. To determine the focus on the object side, the process is reversed. The conjugate line is constructed parallel to the image line. Its intersection with the object line corresponds to the object-space focus.

2.7 Vignetting. Clear Aperture

Vignetting has come to mean the *unwanted* obstruction of a ray by the edge of a lens or by a stop. *Clear aperture* is the term used for the diameter of the opening of a lens element. Any ray incident on a refracting surface at a point whose distance from the axis is less than half the clear aperture is guaranteed to be transmitted. Any ray whose axial distance exceeds this value is doomed to be vignetted.

The precise determination of the clear aperture at each refracting surface is best done at a very late stage in the design process. However, at the first- and third-order level of calculation, such as we are concerned with here, certain *ad hoc* estimates of the clear aperture are useful. Moreover, in the context of the y-ȳ diagram, those estimates are particularly valuable, in that they facilitate selection of systems in which vignetting problems can be forseen and avoided.

The clear aperture at any refracting surface in a lens system can be estimated in the following way. We have seen that any paraxial ray may be determined from a linear combination of two basis rays. These are, for all practical purposes, the two paraxial rays normally traced: the chief ray and the marginal ray. The chief ray defines the edge of the field and the center of the dominant aperture of the lens. The marginal ray defines the center of the field and the edge of the dominant aperture. What then of a ray from the edge of the field that passes through the edge of the aperture? The height of this paraxial ray may be found by adding the height of the paraxial chief ray to the height of the paraxial marginal ray. To account for both edges of the aperture, it is convenient to take the sum of the absolute values of the two quantities, thus

$$y = |y| + |\bar{y}| \quad .$$

On the object plane and all its conjugates $y = 0$, so that $y = |\bar{y}|$; and on the pupil planes and stop planes $\bar{y} = 0$, so that $y = |y|$. These points are easy to find on the y-ȳ diagram. Because the paraxial ray whose height is y is everywhere the sum of y and ȳ, a straight line connecting these points on pupil and conjugate planes on the y-ȳ diagram represents the limit of the boundary ray. The figure obtained is a diamond-like diagram, as shown by the dashed lines in Fig.2.9. To avoid vignetting, every ray must be restricted to the interior of this square. It is clear that any lens that lies wholly within this region is likely to have no vignetting problem. It is certain that any lens design that is represented by a y-ȳ diagram with points external to this region is bound to be plagued by them.

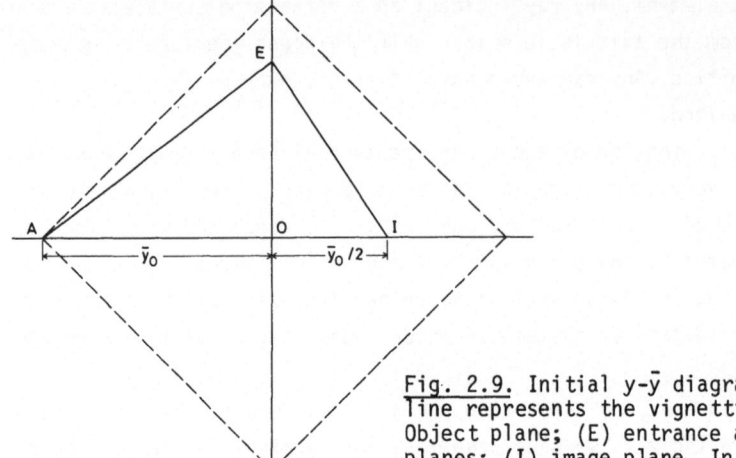

Fig. 2.9. Initial y-ȳ diagram. The dashed
line represents the vignetting diamond. (A)
Object plane; (E) entrance and exit-pupil
planes; (I) image plane. In this diagram, E
also represents the principal planes

2.8 Applications

The y-ȳ diagram provides a way of beginning an optical design. Our description
of the diagram has been to analyze a lens and break it up into its component
elements. In practice the reverse is done. We begin with the y-ȳ diagram itself, and use it to represent the specifications of the lens in graphic terms.
One way to do this is to begin with only the barest sketch of the lens, involving its principal points and its object and image planes. Next the diamond is drawn, representing the vignetting properties of the lens-to-be. The
design process begins with the construction of an open polygon, connecting
the object and image lines in the interior of the diamond. The vertices, representing either subassemblies of the lens or individual refracting surfaces,
can be located in such a way that special requirements on power or diameter
or relative location can be satisfied.

Finally, when the diagram has been completed, the vertices may be represented as refracting or reflecting surfaces. The angle at each vertex can then
be used to estimate the power of the surface it represents, from which the
value of its curvature and the associated refractive indices may be inferred.

The process is highly ambiguous. This has resulted in considerable criti-
cism of the method. However, this ambiguity is perhaps its greatest value.
The diagram represents *unambiguously* the required first-order properties of
the lens. The diagram can be realized in many ways, which are appropriate

first-order designs. Without altering the basic configuration of the diagram, the lens designer can interpret it as any number of distinctly different layouts.

To illustrate these steps, suppose we require a process lens that reduces the size of an object by 50%. Let the diameter of the format on the object plane be denoted by $2\bar{y}_0$. Then the diameter of its image must be $\bar{y}_0$. Let us suppose further that image inversion presents no problem, so that if we take the height of the chief ray on the object plane to be $-\bar{y}_0$ then it must intersect the image plane at a height of $+\bar{y}_0/2$. The first step in constructing the y-$\bar{y}$ diagram is to lay off these distances on the $\bar{y}$ axis, as shown in Fig.2.9. The point $(-\bar{y}_0,0)$, which we will call A, represents the object plane; $(\bar{y}_0/2,0)$, which we denote by I, represents the image plane.

Let us make two more assumptions. First, suppose that the entrance- and exit-pupil planes coincide with the principal planes. Assume next that the radius of the entrance pupil will be equal to $y^\#$. Because the two pupil planes are conjugates with unit magnification, it follows that $y^\#$ must be the diameter of the exit-pupil plane. To represent all of these assumptions, we place the point E (representing both the entrance and exit pupils) on the y axis at the distance $y^\#$ from the origin and draw lines connecting A and E and E and I, shown in Fig.2.10.

Next, let us suppose that the lens is to consist of two refracting elements separated by a stop. Naturally, we would like to keep the diameters of these two units small, and the clearance between the lens and the object and image

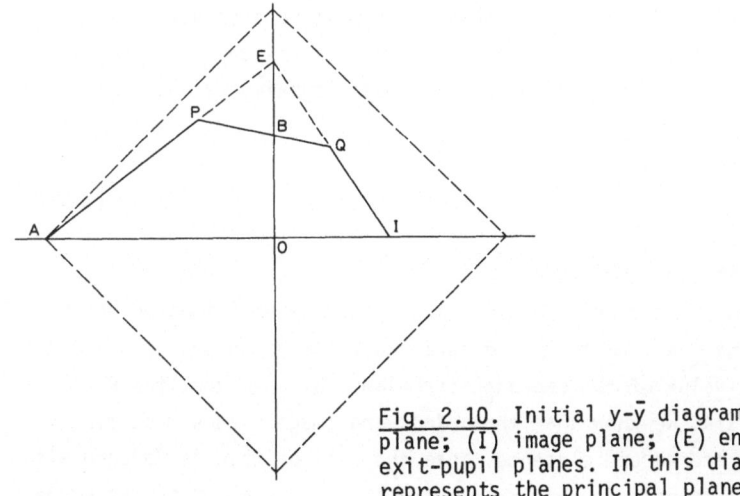

Fig. 2.10. Initial y-$\bar{y}$ diagram. (A) Object plane; (I) image plane; (E) entrance- and exit-pupil planes. In this diagram, E also represents the principal planes

planes as large as possible. These two subassemblies will be represented by
two points on the y-ȳ diagram: P on the line AE and Q on the line EI. We know
that the radius of the first element will be the height y_p of the point P on
the line AE, because y_p is also the height of the marginal paraxial ray at
that lens element. We also know that the distance between the object plane
and the principal point of this lens element denoted by P will be proportional
to the area of the triangle AOP. Similar statements apply to the point Q, its
height y_q, and the triangle QOI. We may locate the points P and Q at any convenient location on the lines AI and EI, respectively. The determination of
a reasonable location of each of these is the next stage in the design process. These points are illustrated in Fig.2.10.

From (2.3.2), we have that $t_A = 2A_A/h$ and $t_I = 2A_I/h$. Here t_A and t_I represent, respectively, the object distance and the image distance. As usual, h
is the Lagrange invariant. The quantities A_A and A_I represent the areas of,
respectively, triangles AOP and QOI, so that

$$A_A = \bar{y}_0 y_p/2 \quad , \quad A_I = (\bar{y}_0/2)y_q/2 \quad .$$

Putting these together gives

$$t_A = \bar{y}_0 y_p/h \quad , \quad t_I = \bar{y}_0 y_q/2h \quad . \tag{2.8.1}$$

At this point, we find that our two requirements are contradictory. Because $\bar{y}_0$ and h are fixed, any increase of t_A and t_I must be accompanied by
an increase of y_p and y_q, signalling an increase of the diameters of the two
subassemblies. This is a stage where trade-offs must be considered.

For the purpose of our example, let us set object distance and image distance equal to the radius of the object and image formats, respectively. Then
$t_A = \bar{y}_0$, $t_I = \bar{y}_0/2$, and

$$y_p = y_q = h \quad , \tag{2.8.2}$$

obtained by substitution into (2.8.1).

Thus, under the assumptions and constraints that we have imposed on the
design, we can conclude from the y-ȳ diagram that the heights of P and Q are
both equal to the value of the Lagrange invariant. Because the line PQ is
parallel to the axis, we know that the object and image planes must lie on
foci of the first and second elements. Note that all of this is deduced without referring to thickness, curvatures, refractive indices, or any other de-

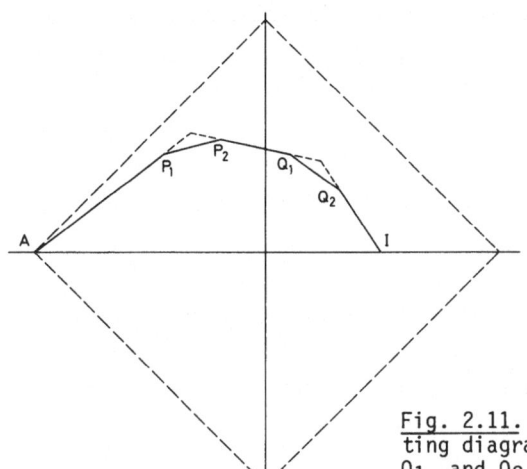

Fig. 2.11. Subassembly detail and vignetting diagram. To avoid vignetting, P_1, P_2,
Q_1, and Q_2 must lie in the interior of the
diamond

sign parameter. Because P and Q are constrained to two known straight lines,
their $\bar{y}$ coordinates may be obtained.

The points P and Q refer to the principal points of the two subsystems.
Refer once again to Fig.2.10. The straight line that connects P and Q intersects the y axis at the point B. This point represents the system's stop,
whose radius y_B must also be the y coordinate of point B. The location of
the stop relative to the principal points of two subassemblies is provided
by the areas of the triangles POB and BOQ.

The next step in the design process is to define P and Q. The procedure
we follow here is just a repetition of what we have already done — subdivide
each element into its constituent components and represent them by additional
points on the y-$\bar{y}$ diagram. In principle, these can again be subdivided *et
cetera, ad infinitum.*

For our purposes, let us take the subassemblies P and Q to be simple lenses.
Each will be a piece of glass with two refracting surfaces. Each surface can
then be represented by a single point on the y-$\bar{y}$ diagram, as shown in Fig.2.11.
Thus, P becomes P_1 and P_2, and Q becomes Q_1 and Q_2. The thickness of the elements is now determined by the triangles P_1OP_2 and Q_1OQ_2. Note that now the
front focal distance and the back focal distances may no longer be exactly
equal to $\bar{y}_0$ and $\bar{y}_0/2$, respectively. These four points may be moved at will,
as long as the areas of the appropriate triangles remain positive.

To insure that vignetting does not occur, construct the diamond-shaped
vignetting diagram as shown in Fig.2.11. Then arrange the points P_1, P_2, Q_1,
and Q_2 so that they all lie in its interior.

Now we must obtain the final thicknesses and curvatures of these elements. This can be done in several ways. Perhaps the easiest way to see this is by means of (2.4.4). By measuring the angle between the lines meeting at, say, point P_1, we can determine the power of the surface represented by that point. By assuming a reasonable value for the refractive index of the glass, it is a simple matter to determine the correct curvature for that surface.

This brief sketch is meant to illustrate an approach to initiating the optical-design process using the y-ȳ diagram. This treatment is meant to be superficial. The point is that the lens designer may interpret the y-ȳ diagram as a number of equivalent first-order designs all of which satisfy the prescribed properties. He may then select the best of these, using his knowledge and experience as a guide. A more descriptive account for optical designers may be found in [2.5] as well as in [2.6], and [2.7,8] also provide descriptions of the process. BESENMATTER [2.9-12] has published several studies of the application of the method to the design of zoom systems.

However, our interest in the subject is concerned with another property of the y-ȳ diagram. With it, we will be able to demonstrate quickly and easily some properties of modules that would be horrendously difficult to prove by more formal methods. For that reason alone, this discussion has been included here.

3. The Two-Surface System

In this chapter, we move one step closer to the precise definition of the module. Here we study the general two-surface system and assemble a collection of formulas and equations for use in subsequent chapters.

3.1 Paraxial Formalities

We have described the module as consisting of two spherical refracting or reflecting surfaces. Here we will formulate the paraxial and third-order properties of a two-surface system preliminary to defining and developing the module concept. As always, the *main focus* is the focus to which all aberration calculations are referred.

Let the curvatures of the two spherical refracting surfaces be denoted by c_1 and c_2. Let these surfaces separate three optical media whose refractive indices are N_0, N_1, and N_2 and whose dispersions are D_0, D_1, and D_2, respectively. Let the axial distance between the two surfaces be t_1. Let the plane of the main focus lie at a distance t_0 from the first surface. These relationships are illustrated in Fig.3.1.

Furthermore, let y_0 be the height of a ray on the plane of the main focus and let u_0 be its slope. We have seen that, from (1.2.7),

$$\begin{pmatrix} y_1 \\ u_1 \end{pmatrix} = \begin{pmatrix} 1 & t_0 \\ -(1-\mu_1)c_1 & \mu_1-(1-\mu_1)c_1 t_0 \end{pmatrix} \begin{pmatrix} y_0 \\ u_0 \end{pmatrix} , \qquad (3.1.1)$$

where $\mu_1 = N_0/N_1$, and where y_1 represents the height of the ray at the first surface and u_1 its slope after refraction at that surface.

From (1.2.5), the formula for transfer, we can see that the point where this refracted ray crosses the axis is given by $0 = y_1 + t u_1$, whence we obtain the formula

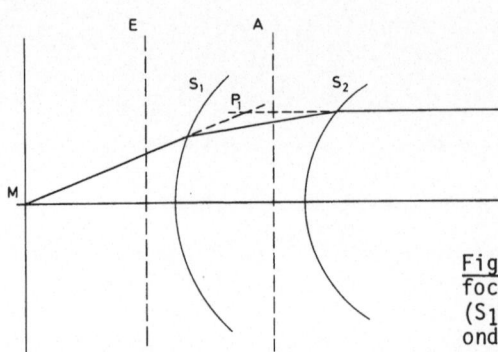

Fig. 3.1. Two-surface system. (M) Main
focus; (P$_1$) first principal point;
(S$_1$) first refracting surface; (S$_2$) second refracting surface; (E) entrance
pupil; (A) exit pupil

$$t_1 = -y_1/u_1 \quad , \tag{3.1.2}$$

where t_1 represents the distance from the first surface to the crossover
point.

The height and slope of this ray, after refraction at the second surface,
is given by a matrix formula identical to (3.1.1),

$$\begin{pmatrix} y_2 \\ u_2 \end{pmatrix} = \begin{pmatrix} 1 & t_1 \\ -(1-\mu_2)c_2 & \mu_2-(1-\mu_2)c_2t_1 \end{pmatrix} \begin{pmatrix} y_1 \\ u_1 \end{pmatrix} \quad ,$$

where $\mu_2 = N_1/N_2$. Substituting for y_1 and u_1 from (3.1.1) results in

$$\begin{pmatrix} y_2 \\ u_2 \end{pmatrix} = \begin{pmatrix} 1 & t_1 \\ -(1-\mu_2)c_2 & \mu_2-(1-\mu_2)c_2t_1 \end{pmatrix} \begin{pmatrix} 1 & t_0 \\ -(1-\mu_1)c_1 & \mu_1-(1-\mu_1)c_1t_0 \end{pmatrix} \begin{pmatrix} y_0 \\ u_0 \end{pmatrix} \quad .$$

By completing the matrix multiplication, we obtain

$$\begin{pmatrix} y_2 \\ u_2 \end{pmatrix} = \left(\begin{array}{c} 1-(1-\mu_1)t_1c_1 \\ -(1-\mu_2)c_2-\mu_2(1-\mu_1)c_1+(1-\mu_2)(1-\mu_1)c_2t_1c_1 \end{array} \right.$$
$$\left. \begin{array}{c} N_0f \\ \mu_2\mu_1-\mu_2(1-\mu_1)c_1t_0-N_0f(1-\mu_2)c_2 \end{array} \right) \begin{pmatrix} y_0 \\ u_0 \end{pmatrix} \quad , \tag{3.1.3}$$

where

$$N_0f = t_0 + \mu_1t_1 - (1-\mu_1)t_1c_1t_0 \quad . \tag{3.1.4}$$

The crossover point for this refracted ray is obtained in exactly the same way as for (3.1.2),

$$t_2 = -y_2/u_2 \quad . \tag{3.1.5}$$

We require that the image of the axial point be at infinity, thus allowing us to call the plane at t_0 the *main focus*. For this to happen, u_2 must vanish whenever $y_0 = 0$. Referring to (3.1.3), we can see that the appropriate equation is

$$\mu_2\mu_1 - \mu_2(1-\mu_1)c_1t_0 - N_0f(1-\mu_2)c_2 = 0 \quad . \tag{3.1.6}$$

The next step is to introduce a new parameter q, defined by

$$q = N_1[\mu_1-(1-\mu_1)c_1t_0] \quad . \tag{3.1.7}$$

We now have three equations, (3.1.4,6,7), which can be considered as a system of simultaneous equations in the unknowns c_2, t_1, and c_1. Solving this system for these quantities leads to

$$c_2 = q/[N_2N_0f(1-\mu_2)] \quad , \tag{3.1.8}$$

$$t_1 = N_1(N_0f-t_0)/q \quad , \tag{3.1.9}$$

and

$$c_1 = (N_0-q)/[N_1t_0(1-\mu_1)] \quad . \tag{3.1.10}$$

Note that c_2, t_1, and c_1 appear as functions of f, q, and t_0 as well as the three refractive indices. Substituting the relations given in (3.1.8-10) into (3.1.2,3) results in

$$\begin{pmatrix} y_1 \\ u_1 \end{pmatrix} = \begin{pmatrix} 1 & t_0 \\ -(N_0-q)/N_1t_0 & q/N_1 \end{pmatrix} \begin{pmatrix} y_0 \\ u_0 \end{pmatrix} \tag{3.1.11}$$

and

$$\begin{pmatrix} y_2 \\ u_2 \end{pmatrix} = \begin{pmatrix} N_0(fq+t_0-N_0f)/t_0q & N_0f \\ -1/N_2f & 0 \end{pmatrix} \begin{pmatrix} y_0 \\ u_0 \end{pmatrix} \quad . \tag{3.1.12}$$

3.2 Cardinal Points

We next examine the cardinal points associated with this two-surface system. We begin by writing the matrix equation for the shift of a reference plane. This turns out to be nothing more than the transfer matrix given in (1.2.6) with the proper interpretation. Let the height of a ray, with slope u on the reference plane, be y. Let the plane be shifted a distance d and let the height of the same ray on the new plane be $y^{\#}$. Then, from (1.2.6), the relationship is given by

$$\begin{pmatrix} y^{\#} \\ u \end{pmatrix} = \begin{pmatrix} 1 & d \\ 0 & 1 \end{pmatrix} \begin{pmatrix} y \\ u \end{pmatrix} . \tag{3.2.1}$$

Referring now to (3.1.12), we can see that a shift of the object plane by an amount d_0 and the image plane by an amount d_2 results in

$$\begin{pmatrix} y_2^{\#} \\ u_2 \end{pmatrix} = \begin{pmatrix} 1 & d_2 \\ 0 & 1 \end{pmatrix} \begin{pmatrix} A & N_0 f \\ -1/N_2 f & 0 \end{pmatrix} \begin{pmatrix} 1 & -d_0 \\ 0 & 1 \end{pmatrix} \begin{pmatrix} y_0^{\#} \\ u_0 \end{pmatrix} ,$$

where

$$A = N_0 [t_0 - f(N_0 - q)]/t_0 q . \tag{3.2.2}$$

By carrying out the multiplication, we obtain

$$\begin{pmatrix} y_2^{\#} \\ u_2 \end{pmatrix} = \begin{pmatrix} A - d_2/N_2 f & N_0 f - d_0 A + d_0 d_2/N_2 f \\ -1/N_2 f & d_0/N_2 f \end{pmatrix} \begin{pmatrix} y_0^{\#} \\ u_0 \end{pmatrix} . \tag{3.2.3}$$

We already have one of the cardinal points, which we have called the main focus. The second focus is found by setting $u_0 = 0$ in (3.2.3), then finding a value of d_2 for which $y_2^{\#} = (A - d_2/N_2 f) y_0^{\#}$ equals zero. Setting the quantity in parentheses equal to zero, solving for d_2, and making use of (3.2.2), we obtain F_2, the distance from the second surface to the second focus,

$$F_2 = d_2 = N_2 N_0 f [t_0 - f(N_0 - q)]/t_0 q . \tag{3.2.4}$$

The remaining cardinal points come in conjugate pairs. We need first of all a relationship that assures that y_2 depends only on y_0 and is therefore independent of u_0. This occurs when, referring to (3.2.3),

$$N_0 f - d_0 A + d_0 d_2 / N_2 f = 0 \quad .$$

When this is solved for d_2, we get

$$d_2 = \frac{N_2 N_0 f}{d_0} \left[\frac{d_0}{t_0 q} (fq + t_0 - N_0 f) - f \right] \quad , \tag{3.2.5}$$

the condition that assures that a point on a plane located at d_0 is imaged onto a plane located at d_2. We must keep in mind that d_0 is measured from the main focus and that d_2 is measured from the second surface.

The relationship between object and image plane is now

$$y_2^\# = y_0^\# N_0 f / d_0 \quad . \tag{3.2.6}$$

The principal planes constitute the unique pair of conjugate planes for which the magnification is unity. This occurs when $d_0 = N_0 f$. Substituting this into (3.2.5) results in $d_2 = N_2 f (1-A)$.

The front focal length is the distance between the main focus and the corresponding principal plane. This turns out to be exactly equal to $N_0 f$. The rear focal length is the distance between the second focus and the second principal plane. This turns out to be equal to $d_2 - F_2 = -N_2 f$, as expected. We conclude that the power of the system is exactly $1/f$.

The nodal points are determined in a manner similar to the method used to find the principal points. The front and rear nodal points are defined as the unique pair of conjugate points that have the property that any ray passing through one emerges from the other undeviated. Because we are dealing with conjugate pairs, (3.2.5) holds. The nodal points clearly must lie on the axis, so that $y_0^\#$ and $y_2^\#$ must both be zero. We then obtain from (3.2.3)

$$u_2 = u_0 d_0 / N_2 f \quad .$$

Now u_2 must equal u_0. This leads us to

$$d_0 = N_2 f \quad .$$

When this is substituted into (3.2.5), we get

$$d_2 = \frac{N_0 f}{t_0 q} \ [N_2(fq+t_0-N_0 f)-t_0 q] \quad .$$

Summarizing these results, we have for the principal points

$$d_0 = N_0 f \quad , \quad d_2 = \frac{N_2 f}{t_0 q} \ (N_0-q)(N_0 f-t_0) \quad , \tag{3.2.7}$$

and for the nodal points,

$$d_0 = N_2 f \quad , \quad d_2 = \frac{N_0 f}{t_0 q} \ [N_2(fq+t_0-N_0 f)-t_0 q] \quad . \tag{3.2.8}$$

3.3 Third-Order Image Errors

We now distinguish two paraxial rays; the marginal ray, which will always pass through the main focus, and the chief ray, whose properties will for the moment be unspecified. Following a common convention, marginal-ray quantities will be unaccented, while chief-ray quantities will be distinguished by a superior bar. Thus y and u denote the height and slope of the marginal ray, respectively, whereas $\bar{y}$ and $\bar{u}$ refer to similar quantities for the chief ray. In the two-surface system being considered here, we have set $\bar{y}_0 = u_2 = 0$.

First of all. we need to calculate the four refraction invariants given in (1.2.4) for the marginal and chief rays at each of the two surfaces. Here we must refer to (3.1.8-10) to obtain

$$a_1 = u_0 \mu_1 (N_1-1)/(1-\mu_1) \quad , \quad \bar{a}_1 = \bar{u}_0 \mu_1 (N_1-1)/(1-\mu_1)$$
$$+ \bar{y}_0 \mu_1 (N_0-q)/(1-\mu_1) t_0 \quad ,$$

$$a_2 = u_0 Q/(1-\mu_2) \quad , \quad \bar{a}_2 = \bar{u}_0 q/(1-\mu_2)$$
$$+ \bar{y}_0 [\mu_2 t_0 - f(N_0-q)]/f t_0 (1-\mu_2) \quad . \tag{3.3.1}$$

Next we refer to (1.2.11) and calculate s and $\bar{s}$ at each of the two surfaces. Again using (3.1.8-10) as well as (1.2.11), we get

$$s_1 = -u_0 (N_1^2-N_0 q)/N_1^2 N_0 \quad , \quad \bar{s}_1 = -\bar{u}_0 (N_1^2-N_0 q)/N_1^2 t_0 - \bar{y}_0 (N_0-q)/N_1^2 t_0 \quad ,$$

$$s_2 = -u_0 q/N_1^2 \quad , \quad \bar{s}_2 = -\bar{u}_0 q/N_1^2 + \bar{y}[f(N_0-q)-\mu_2^2 t_0]/N_1^2 f t_0 \quad .$$
$$\tag{3.3.2}$$

We are now in a position to calculate the third-order image-error coefficients. Rewrite (3.3.1,2) as

$$a_1 = u_0 \alpha_1 \quad , \quad \bar{a}_1 = \bar{u}_0 \alpha_1 + \bar{y}_0 \beta_1 \quad ,$$
$$a_2 = u_0 \alpha_2 \quad , \quad \bar{a}_2 = \bar{u}_0 \alpha_2 + \bar{y}_0 \beta_2 \quad , \tag{3.3.3}$$

where

$$\alpha_1 = \mu_1 (N_1 - q)/(1 - \mu_1) \quad , \quad \beta_1 = \mu_1 (N_0 - q)/(1 - \mu_1) t_0 \quad ,$$
$$\alpha_2 = q/(1 - \mu_2) \quad\quad\quad , \quad \beta_2 = [\mu_2 t_0 - f(N_0 - q)]/(1 - \mu_2) t_0 f \quad , \tag{3.3.4}$$

and

$$s_1 = -u_0 \gamma_1 \quad , \quad\quad\quad\quad \bar{s}_1 = -\bar{u}_0 \gamma_1 - \bar{y}_0 \delta_1 \quad ,$$
$$s_2 = -u_0 \gamma_2 \quad , \quad\quad\quad\quad \bar{s}_2 = -\bar{u}_0 \gamma_2 - \bar{y}_0 \delta_2 \quad , \tag{3.3.5}$$

where

$$\gamma_1 = (N_1^2 - N_0 q)/N_1^2 N_0 \quad , \quad \delta_1 = (N_0 - q)/N_1^2 t_0 \quad ,$$
$$\gamma_2 = q/N_1^2 \quad\quad\quad\quad\quad , \quad \delta_2 = -[f(N_0 - q) - \mu_2^2 t_0]/N_1^2 f t_0 \quad . \tag{3.3.6}$$

In addition, we recall from (3.1.11,12) that

$$y_1 = u_0 t_0 \quad , \quad \bar{y}_1 = \bar{u}_0 t_0 + \bar{y}_0 \quad ,$$
$$y_2 = u_0 N_0 f \quad , \quad \bar{y}_2 = \bar{u}_0 N_0 f + \bar{y}_0 A \quad , \tag{3.3.7}$$

where A is given by (3.2.2).

Substituting these relations into (1.2.12-14), we obtain expressions for spherical aberration

$$b = u_0^4 U_1 \quad , \tag{3.3.8}$$

for coma

$$\mathit{6} = u_0^3 (\bar{u}_0 U_1 + \bar{y}_0 U_2) \quad , \tag{3.3.9}$$

and for astigmatism

$$c = u_0^2(\bar{u}_0 U_1 + 2\bar{u}_0\bar{y}_0 U_2 + \bar{y}_0^2 U_3) \quad . \tag{3.3.10}$$

In these three expressions,

$$U_1 = t_0\gamma_1\alpha_1^2 + N_0 f\gamma_2\alpha_2^2 \quad ,$$

$$U_2 = t_0\gamma_1\alpha_1\beta_1 + N_0 f\gamma_2\alpha_2\beta_2 \quad ,$$

$$U_3 = t_0\gamma_1\beta_1^2 + N_0 f\gamma_2\beta_2^2 \quad . \tag{3.3.11}$$

3.4 Third-Order Pupil Errors

The equations for the pupil errors, as can be seen from (1.2.24-26), are identical to those of the image errors except that all unaccented quantities are replaced by accented quantities and vice versa. The result of this rather simple operation is anything but simple, as the following collection of formulas shows.

We will reverse the usual order and begin with pupil astigmatism, from (1.2.26), and (3.3.3-7); we obtain

$$\bar{c} = u_0^2\left[\bar{u}_0^2 U_1 + \bar{u}_0\bar{y}_0(W_1 + W_2) + \bar{y}_0^2 W_3\right] \quad , \tag{3.4.1}$$

where U_1 can be found in (3.3.11), and where the other new quantities are given by

$$W_1 = t_0\alpha_1^2\delta_1 + N_0 f\alpha_2^2\delta_2 \quad ,$$

$$W_2 = \alpha_1^2\gamma_1 + \alpha_2^2 A\gamma_2 \quad ,$$

$$W_3 = \alpha_1^2\delta_1 + \alpha_2^2 A\delta_2 \quad . \tag{3.4.2}$$

The quantity A is defined in (3.2.2). Next comes pupil coma, in which we use (1.2.25),

$$\bar{\delta} = u_0\left[\bar{u}_0^3 U_1 + \bar{u}_0^2\bar{y}_0(W_1 + W_2 + U_2) + \bar{u}_0\bar{y}_0^2(W_3 + V_1 + V_2) + \bar{y}_0^3 V_3\right] \quad . \tag{3.4.3}$$

The new quantities introduced in this equation are defined by

$$V_1 = t_0\alpha_1\beta_1\delta_1 + N_0f\alpha_2\beta_2\delta_2 \quad ,$$

$$V_2 = \alpha_1\beta_1\gamma_1 + \alpha_2\beta_2 A\gamma_2 \quad ,$$

$$V_3 = \alpha_1\beta_1\delta_1 + \alpha_2\beta_2 A\delta_2 \quad . \tag{3.4.4}$$

Finally, we treat pupil spherical aberration, which derives from (1.2.24):

$$\overline{6} = \overline{u}_0^4 U_1 + \overline{u}_0^3\overline{y}_0(W_1+W_2+2U_2) + \overline{u}_0^2\overline{y}_0^2(W_3+2V_1+2V_2+U_3)$$

$$+ \overline{u}_0\overline{y}_0^3(2V_3+V_4+V_5) + \overline{y}_0^4 V_6 \quad , \tag{3.4.5}$$

where

$$V_4 = t_0\beta_1^2\delta_1 + N_0f\beta_2^2\delta_2 \quad ,$$

$$V_5 = \beta_1^2\gamma_1 + \beta_2^2 A\gamma_2 \quad ,$$

$$V_6 = \beta_1^2\delta_1 + \beta_2^2 A\delta_2 \quad . \tag{3.4.6}$$

3.5 Third-Order Field Aberrations

We have been examining the group of third-order aberrations that are frequently referred to as the image errors — spherical aberration, coma, and astigmatism. These are concerned with the defect of the image of a single object point; we have seen that by using modules we have absolute control over two of them.

Now we turn to the so-called field aberrations, which related to the structure and the defects of the image of an extended object. These are field curvature and distortion.

Two components enter into the calculation of third-order field curvature, third-order astigmatism and the Petzval contribution. However, astigmatism will be made identically zero. All we will need to know about field curvature is contained in the Petzval contribution alone, which is given by

$$p = [N_2f(N_0-q)+t_0q]/N_2N_1N_0ft_0 \quad . \tag{3.5.1}$$

This is obtained from (1.2.15) by use of (3.1.8,10).

We now turn to calculation of image distortion and use (1.2.28),

$$e_j = \bar{f}_j - h\Delta(\bar{u}_j^2) \quad,$$

where $\bar{f}_j$ is pupil coma and h is the Lagrange invariant. In the notation of this chapter, the contribution to distortion from the two surfaces is

$$e = \bar{\delta} - h\left(\bar{u}_2^2 - \bar{u}_0^2\right) \quad.$$

Making use of (3.1.12), we obtain

$$e = \bar{\delta} - h\left(\bar{y}_0^2 - N_2^2 f^2 u_0^2\right) / N_2^2 f^2 \quad. \tag{3.5.2}$$

For pupil distortion, we use a different battery of calculations. Beginning with (1.2.23), we find

$$\bar{e}_j = f_j + h(u_j^2) \quad,$$

where f_j is image coma. The contribution from these two surfaces is given by

$$\bar{e} = \delta + hu_0^2 \quad, \tag{3.5.3}$$

because $u_2 = 0$ by (3.1.12).

3.6 Primary Chromatic Aberrations

The calculation of the primary chromatic aberrations proceeds in very much the same way. We begin with (1.3.1-3), into which we substitute (3.3.3,4) and so obtain the longitudinal chromatic aberration,

$$\begin{aligned}
\ell &= a_1 y_1 \Delta(D_0/N_0) + a_2 y_2 \Delta(D_1/N_1) \\
&= u_0^2 [\alpha_1 t_0 (D_1/N_1 - D_0/N_0) + \alpha_2 N_0 f (D_2/N_2 - D_1/N_1)] \\
&= u_0^2 L \quad, \tag{3.6.1}
\end{aligned}$$

where

$$\hat{L} = \frac{\mu_1(N_1-q)t_0}{1-\mu_1}\left(\frac{D_1}{N_1}-\frac{D_0}{N_0}\right) + \frac{qN_0f}{1-\mu_2}\left(\frac{D_2}{N_2}-\frac{D_1}{N_1}\right)$$

$$\acute{L} = \alpha_1t_0\left(\frac{D_1}{N_1}-\frac{D_0}{N_0}\right) + N_0f\alpha_2\left(\frac{D_2}{N_2}-\frac{D_1}{N_1}\right) \quad . \tag{3.6.2}$$

We also obtain the transverse chromatic aberration,

$$\mathit{t} = a_1y_1(D_0/N_0) + a_2y_2(D_1/N_1)$$

$$= u_0[t_0(u_0\alpha_1+y_0\beta_1)(D_1/N_1-D_0/N_0)+N_0f(u_0\alpha_2y_0\beta_2)(D_2/N_2-D_1/N_1)]$$

$$= u_0\bar{u}_0\acute{L} + u_0\bar{y}_0\acute{T} \quad , \tag{3.6.3}$$

where

$$\hat{T} = \frac{\mu_1(N_0-q)}{1-\mu_1}\left(\frac{D_1}{N_1}-\frac{D_0}{N_0}\right) + \frac{\mu_2t_0-f(N_0-q)}{(1-\mu_2)t_0} N_0\left(\frac{D_2}{N_2}-\frac{D_1}{N_1}\right) \quad ,$$

$$\acute{T} = \beta_1t_0\left(\frac{D_1}{N_1}-\frac{D_0}{N_0}\right) + \beta_2N_0f\left(\frac{D_2}{N_2}-\frac{D_1}{N_1}\right) \quad .$$

This concludes this phase of the calculations.

3.7 The y-ȳ Diagram

The next item on the agenda is the construction of a y-ȳ diagram to represent the two-surface system described in Sect.3.1 and shown in Fig.3.1. The first step in the construction is to regard the system as a simple input—output device. Figure 3.2 shows the y-ȳ diagram as consisting of a point A on the ȳ axis, that represents the object plane. A line segment connects A to a second point P which represents the object and image principal planes. From P, a line parallel to the ȳ axis indicates that the image point is at infinity and, therefore, that the marginal ray that exits from the lens is parallel to the lens axis.

The point A is situated on the ȳ axis at a distance $\bar{y}_0$ to the left of the origin. This shows that the height of the chief ray on the object plane of the lens is equal to $-\bar{y}_0$, as indicated in Fig.3.1. From (3.1.12), we see that

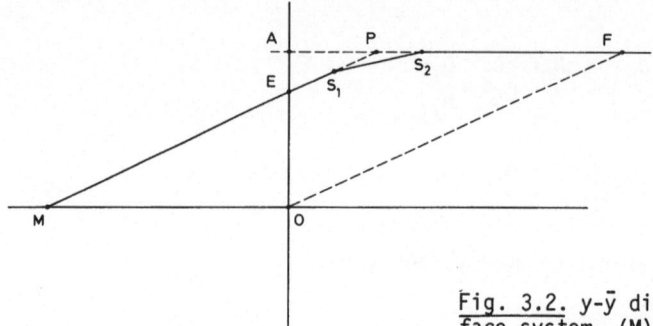

Fig. 3.2. y-ȳ diagram for the two-surface system. (M) Main focus; (P) object
and image principal planes; (F) second
focal plane; (S$_1$) first refracting surface; (S$_2$) second refracting surface;
(E) entrance pupil; (A) exit pupil

the height of the marginal ray on the second surface is $y_2 = N_0 f u_0$. It follows
that this is also the height of the y coordinate of the point P.

We have seen that the front focal length of the two-surface system is
equal to $N_0 f$ and that this is the distance from the main focus to the object
principal point. Then the area of the triangle AOP is related to the front
focal length by (2.3.2). Note that the area of this triangle does not depend
on the location of the point P.

The second focus of the two-surface system is found by constructing a line
through the origin parallel to AP. Its intersection with the horizontal line,
marked F in Fig.3.2, represents the second focus. The area of the triangle
POF is related to the back focal length $N_2 f$ by (2.3.2).

The slope of the conjugate line OP is given by (2.6.2). The slope of the
conjugate line to the representations of nodal points, not shown in Fig.3.2,
is given by (2.6.3).

Finally, we come to the points that represent the two refracting surfaces.
The first of these, denoted S$_1$ in Fig.3.2, must lie on the line AP. The point
that represents the second refracting surface, S$_2$, must lie on the line PF.
The y-ȳ diagram is completed by drawing a line connecting these points.

4. The Module

In the previous chapter, we obtained the properties of a special two-surface optical system. We calculated certain of the Seidel aberrations and the primary chromatic aberrations of this system associated with its main focus. The next step in the development of the module is to determine the relationship between refractive indices, curvatures, thicknesses, and focal distances for which third-order spherical aberrations is identically zero. Then we will obtain a relationship between the height and slope of the chief ray that will result in zero third-order astigmatism. This will lead, in turn, to the definition of pupil planes in the three media for which third-order astigmatism is identically zero.

4.1 Spherical Aberration

We begin with (3.3.8), the expression for the sum of the coefficients of third-order spherical aberration for the two-surface system described in the last chapter. Setting this expression equal to zero, we obtain, from (3.3.12),

$$U_1 = t_0 \gamma_1 \alpha_1^2 + N_0 f \gamma_2 \alpha_2^2 = 0 \quad .$$

Substituting for the α's and γ's from (3.3.4,6) and clearing fractions, we obtain

$$t_0 (N_2 - N_1)^2 (N_1 - q)^2 (N_1^2 - N_0 q) + N_2^2 f (N_1 - N_0)^2 q^3 = 0 \quad . \tag{4.1.1}$$

In the substitution from (3.3.4), we use the definition of μ_i given in (1.2.3), so that $\mu_1 = N_0/N_1$, $\mu_2 = N_1/N_2$. Equation (4.1.1) is expanded and terms in the powers of q are collected to obtain a cubic polynomial in q,

$$\left[N_2^2 f(N_1-N_0)^2 - N_0 t_0 (N_2-N_1)^2\right]q^3$$
$$+ t_0 N_1 (N_2-N_1)^2 \left[(N_1+2N_0)q^2 - N_1(2N_1+N_0)q + N_1^3\right] = 0 \quad . \tag{4.1.2}$$

Clearly, if we can solve this equation for q as a function of t_0, f, and the refractive indices, we will establish a relationship between these quantities that will assure that third-order spherical aberration is identically zero.

4.2 The Cubic Polynomial

Quite apart from its applications here, the cubic and its solution is part of a most fascinating historical development. In the sixteenth century, algebraic-problem solving was a popular competitive sport, perhaps a little like twentieth-century tennis. The presentation of a solution to a specific numerical problem was a public display, a kind of ritual that suited the flamboyant personalities of that day. Prizes were awarded to the victors at the nearest bistro, where ridicule and other forms of abuse were heaped on the heads of the unfortunate losers.

The first step toward a general solution of the cubic was made by Scipio Del Ferro in about 1520. However, he refused to part with his secret and carried it to his grave, to which he was perhaps hastened by too long a string of victory celebrations. His papers came into the possession of his son-in-law, Annibale della Nave as well as Antonio Maroa Fiore. Evidently, they chose to keep whatever they knew of the method secret, preferring to display their virtuosity in solving cubics in public competition, thus bolstering their egos and lining their purses.

The next member of the cast of this little drama is Niccolo Tartaglia (1499-1557). "Tartaglia", meaning "stutterer" or "stammerer", was in reality only a nickname. As a small child, he was caught in a French massacre of the population of Brescia, in which he was severely wounded in the face and mouth, which left him with a permanent speech defect. However, this did not deter him from taking part, quite successfully, in the public mathematical competitions. In these he was outstanding and was raised to considerable prominence in the mathematical community; as a result, he was challenged frequently by other mathematicians who thought they could best him. Among these was Fiore.

The challenge was made and accepted, a date was set; thirty problems were exchanged, the wager on each being a banquet.

But Fiore had but one string to his fiddle, a particular form of the cubic polynomial. All thirty of his problems were of this type. Tartaglia, on the other hand, was far broader in his abilities and presented Fiore with a far more diverse range of problems. Tartaglia worked frantically and desperately. Finally, on the night of 12 February, 1535, he saw the trick. Because all of Fiore's problems were of the same type, once he was able to solve one of them, all the rest came easily. At the public exhibition he won all thirty banquets, hands down.

Shortly afterwards Gerolamo Cardano (1501-1576) entered the picture. He was a physician (he was called on to treat the archbishop of Scotland), astrologer (he forecast a long and prosperous life for the young King Edward VI of England), gambler (he wrote a gambler's handbook), and a mathematician of considerable ability and renown. He was in the process of writing a book on algebra, had heard of Tartaglia's great success, and intended to include a description of it in his opus. He sent several emmissaries to persuade Tartaglia to reveal it. Tartaglia refused. Cardano attempted to discover the rule on his own, but was unable to do so. Several years later, when his book was being prepared for publication, Cardano tried again through yet another agent. Tartaglia again refused, saying that he preferred to publish his results in a book of his own when the time was right. Nevertheless, Cardano, through his agent and through a correspondence, maintained continuous pressure on Tartaglia. Finally, Cardano persuaded Tartaglia to visit him at Milano. There Cardano swore an oath not to reveal the secret; indeed, he promised to write the formula in cypher so that should he die his heirs would be unable to discover Tartaglia's method. Tartaglia finally gave in and turned over the formula in the form of a poem, and a rather poor one at that.

Cardano was able to generalize Tartaglia's method. This led him into a long excursion into the realm of what we now understand to be complex numbers. During this period his student, Lodovico Ferrari (1522-1565), a collaborator and confidant, discovered the method of solving the quartic equation, which depends on a knowledge of the solution of a particular cubic.

Tartaglia still failed to publish. Yet Tartaglia's work on the cubic was the cornerstone on which the further work of Cardano and Ferrari was erected. There was no way in which their work could be published without revealing the secret Cardano had solemnly sworn not to reveal. The motivation was too strong. The immovable object eventually gave way to the irresistible force and Ferrari's petulant urgings. The great oath was abrogated and Tartaglia's method was published by CARDANO [4.1] in 1545, in a new book, *Ars Magna*. See also ECKMANN [4.2], ORE [4.3], SMITH [4.4], GHERARDI [4.5] and, of course, TARTAGLIA [4.6].

Our work on the cubic must begin with some well-known facts on the quadratic

$$Ax^2 + Bx + C = 0 \quad,$$

whose two solutions are provided by the formula

$$x_\pm = [-B \pm (B^2 - 4AC)^{1/2}]/2A \quad . \tag{4.2.1}$$

We note that this can be modified in any number of ways. For example, the discriminant can be factored,

$$B^2 - 4AC = (B + 2\sqrt{AC})(B - 2\sqrt{AC}) \quad,$$

so that, by multiplying the numerator and denominator of (4.2.1) by 2 we obtain

$$x_\pm = -[2B \mp 2(B + 2\sqrt{AC})^{1/2}(B - 2\sqrt{AC})^{1/2}]/4A \quad,$$

which can be written in the form

$$x_\pm = -[(B + 2\sqrt{AC})^{1/2} \mp (B - 2\sqrt{AC})^{1/2}]^2/4A \quad .$$

Next we note that

$$[(B + 2\sqrt{AC})^{1/2} \pm (B - 2\sqrt{AC})^{1/2}][(B + 2\sqrt{AC})^{1/2} \mp (B - 2\sqrt{AC})^{1/2}] = 4\sqrt{AC} \quad,$$

so that

$$x_\pm = -\left(\frac{C}{A}\right)^{1/2} \frac{(B + 2\sqrt{AC})^{1/2} \mp (B - 2\sqrt{AC})^{1/2}}{(B + 2\sqrt{AC})^{1/2} \pm (B - 2\sqrt{AC})^{1/2}} \quad . \tag{4.2.2}$$

Note that if $C = A$, then $x_- = 1/x_+$.

Now consider the general cubic polynomial (see VILHEM [4.7]). With no loss of generality, we may set the leading coefficient equal to unity,

$$x^3 + ax^2 + bx + c = 0 \quad . \tag{4.2.3}$$

The first step in obtaining its solution is to apply the transformation

$$x = y - a/3 \, , \tag{4.2.4}$$

which, when substituted into (4.2.3), yields the so-called reduced cubic,

$$y^3 + (b-a^2/3)y + (c-ab/3+2a^3/27) = 0 \, ,$$

in which the quadratic term does not appear. It is most convenient to rewrite this equation in the form

$$(3y)^3 - 3A(3y) + B = 0 \, , \quad A = a^2 - 3b \, , \quad B = 27c - 9ab + 2a^3 \, . \tag{4.2.5}$$

There are two cases to be considered, determined by the sign of A. When A is positive, set

$$3y = \sqrt{A}(z+1/z) \, , \tag{4.2.6}$$

where z is now the new independent variable; substitute it into (4.2.5) and obtain

$$A^{3/2}(z^3+1/z^3) + B = 0 \, ,$$

which then yields the quadratic in z^3,

$$A^{3/2}z^6 + Bz^3 + A^{3/2} = 0 \, ,$$

whose solution, using (4.2.1), is

$$z^3 = [-B\pm(B^2-4AC)^{\frac{1}{2}}]/2A^{3/2} \, .$$

We have seen, from (4.2.2), that this expression can be written in the form,

$$z^3 = - \frac{(B+2A^{3/2})^{\frac{1}{2}} - (B-2A^{3/2})^{\frac{1}{2}}}{(B+2A^{3/2})^{\frac{1}{2}} + (B-2A^{3/2})^{\frac{1}{2}}} \, . \tag{4.2.7}$$

Here we have dropped the sign ambiguity, because z and its inverse enter the expression for y in (4.2.6) as a sum.

Finally, we extract the cube root and obtain the solution,

$$z_r = -\omega^r \left(\frac{(B+2A^{3/2})^{\frac{1}{2}} - (B-2A^{3/2})^{\frac{1}{2}}}{(B+2A^{3/2})^{\frac{1}{2}} + (B-2A^{3/2})^{\frac{1}{2}}} \right)^{1/3} \quad , \quad r = -1,0,1 \quad , \tag{4.2.8}$$

where $\omega = e^{2\pi i/3}$ is a cube root of unity.

The second case arises when A is negative and parallels the first. For convenience, set $\tilde{A} = -A = 3b - a^2$, so that the reduced cubic becomes

$$(3y)^3 + 3\tilde{A}(3y) + B = 0 \quad .$$

Next, we set

$$3y = \tilde{A}^{\frac{1}{2}}(z-1/z) \quad , \tag{4.2.9}$$

from which, when it is substituted into the reduced cubic, we obtain

$$\tilde{A}^{3/2}(z^3-1/z^3) + B = 0 \quad .$$

Proceeding exactly as before, using (4.2.1), we obtain

$$z^3 = [-B \pm (B^2+4\tilde{A}^3)^{\frac{1}{2}}]/2\tilde{A}^{3/2} \tag{4.2.10}$$

or, from (4.2.2),

$$z^3 = -i \frac{(B+2i\tilde{A}^{3/2})^{\frac{1}{2}} - (B-2i\tilde{A}^{3/2})^{\frac{1}{2}}}{(B+2i\tilde{A}^{3/2})^{\frac{1}{2}} + (B-2i\tilde{A}^{3/2})^{\frac{1}{2}}} \quad , \tag{4.2.11}$$

a very complex expression. As in the previous case, the ambiguity of sign is dropped because, in the equation for y, (4.2.9), z appears only as a difference between z and its reciprocal.

Thus, either (4.2.8) or (4.2.12) provides three values of z which, when substituted into (4.2.4), yield three values of y. When these are inserted into (4.2.4), the three roots of the cubic polynomial, (4.2.3), are obtained:

$$z_r = i\omega^r \left(\frac{(B+2i\tilde{A}^{3/2})^{\frac{1}{2}} - (B-2i\tilde{A}^{3/2})^{\frac{1}{2}}}{(B+2i\tilde{A}^{3/2})^{\frac{1}{2}} + (B-2i\tilde{A}^{3/2})^{\frac{1}{2}}} \right)^{1/3} \quad , \quad r = -1,0,1 \quad . \tag{4.2.12}$$

Here again ω is a complex cube root of unity.

4.3 The Module Cubic

We are now in a position to deal with the cubic polynomial in (4.1.2), which was obtained by setting the expression for third-order spherical aberration equal to zero. Experience dictates that we replace q by a reciprocal quantity, say,

$$q = 1/p \quad , \tag{4.3.1}$$

so that (4.1.2) becomes

$$t_0 N_1 (N_2 - N_1)^2 \left[N_1^3 p^3 - N_1 (2N_1 + N_0) p^2 + (N_1 + 2N_0) p \right]$$
$$+ \left[N_2^2 f (N_1 - N_0)^2 - N_0 t_0 (N_2 - N_1)^2 \right] = 0 \quad , \tag{4.3.2}$$

which is a little easier to manipulate.

From this, we obtain the reduced cubic by applying the substitution prescribed in (4.2.4),

$$p = y + (2N_1 + N_0)/3N_1^2 \quad . \tag{4.3.3}$$

The reduced cubic turns out to be in the form

$$27N_1^6 (N_2 - N_1)^2 t_0 y^3 - 9N_1^2 (N_2 - N_1)^2 (N_1 - N_0)^2 t_0 y$$
$$+ 2(N_2 - N_1)^2 (N_1 - N_0)^3 t_0 + 27N_2^2 N_1^2 f (N_1 - N_0)^2 = 0 \quad ,$$

which is more convenient to write as

$$t_0 (N_2 - N_1)^2 (3N_1^2 y)^3 - 3t_0 (N_2 - N_1)^2 (N_1 - N_0)^2 (3N_1^2 y)$$
$$+ 2t_0 (N_2 - N_1)^2 (N_1 - N_0)^3 + 27N_2^2 N_1^2 f (N_1 - N_0)^2 = 0 \quad . \tag{4.3.4}$$

This is clearly in the form of (4.2.5), where

$$A = (N_1 - N_0)^2 \quad ,$$

$$B = A \left[2t_0 (N_2 - N_1)^2 (N_1 - N_0) + 27N_2^2 N_1^2 f \right] / t_0 (N_2 - N_1)^2 \quad . \tag{4.3.5}$$

The coefficient A is clearly always positive, so that the first case applies. Making use of (3.2.6), we get

$$3N_1^2 y = (N_1 - N_0)(z + 1/z) \quad . \tag{4.3.6}$$

This in turn leads us to (4.2.7), which, after considerable reduction, becomes

$$z^3 = - \frac{\left[27 N_2^2 N_1^2 f + 4 t_0 (N_2 - N_1)^2 (N_1 - N_0)\right]^{\frac{1}{2}} - 3 N_2 N_1 (3f)^{\frac{1}{2}}}{\left[27 N_2^2 N_1^2 f + 4 t_0 (N_2 - N_1)^2 (N_1 - N_0)\right]^{\frac{1}{2}} + 3 N_2 N_1 (3f)^{\frac{1}{2}}} \quad . \tag{4.3.7}$$

This provides the solution of the cubic, (4.1.2). However, it is a formidable monster and needs to be simplified. To do this, we first define a new parameter k:

$$27 N_2^2 N_1^2 k^2 f = 27 N_2^2 N_1^2 f + 4 t_0 (N_2 - N_1)^2 (N_1 - N_0) \quad . \tag{4.3.8}$$

This we solve for t_0, and obtain

$$t_0 = \frac{27 N_2^2 N_1^2 f (k^2 - 1)}{4 (N_2 - N_1)^2 (N_1 - N_0)} \quad . \tag{4.3.9}$$

Finally, we substitute (4.3.8) into (4.3.7) to obtain, after considerable reduction, $z^3 = -(k-1)/(k+1)$, which at least yields

$$z_r = -\omega^{-r} [(k-1)/(k+1)]^{1/3} \quad , \quad r = -1, 0, 1 \quad , \tag{4.3.10}$$

where ω is a cube root of unity. When this is substituted into (4.3.6), we obtain three values for y, which provide three values of p when inserted into (4.3.3). These, in turn, give us three values of q from (4.3.1), the three roots of the cubic polynomial in (4.1.2), exactly as advertised. These are

$$q_r = \frac{3 N_1^2}{(2 N_1 + N_0) - (N_1 - N_0)\left[\omega^r \left(\frac{k-1}{k+1}\right)^{1/3} + \omega^{-r} \left(\frac{k+1}{k-1}\right)^{1/3}\right]} \quad , \quad r = -1, 0, 1 \quad . \tag{4.3.11}$$

There is a point here that sometimes causes confusion. Note that (4.1.1) or (4.1.2), considered as a polynomial in q, has coefficients that depend on t_0. Thus, any solution q of this cubic polynomial must depend functionally on t_0. However, in the solution just obtained, a parameter k was introduced in (4.3.8) upon which depend both q and t_0. We have, therefore, a one-parameter family of values of q and t_0 defined by (4.3.9) and (4.3.11). If the parameter k is eliminated between these two equations, we obtain q as a very complicated function of t_0.

Two cases occur. The first occurs when k is real; it is therefore referred to as the *real case*. For convenience, set

$$\psi^3 = (k-1)/(k+1) \quad . \tag{4.3.12}$$

the inverse relation is, of course,

$$k = (1+\psi^3)/(1-\psi^3) \quad . \tag{4.3.13}$$

Substituting this into (4.3.11) gives

$$q_r = 3N_1^2/\left[(2N_1+N_0)-(N_1-N_0)(\omega^r\psi+\omega^{-r}\psi^{-1})\right] \quad , \quad r = 0,1,2 \quad .$$

Clearly, q_r is real only if $r = 0$. Setting $r = 0$, dropping the subscript, and doing a small amount of manipulating, we get

$$q = -3N_1^2\psi/\left[(N_1-N_0)(\psi^2+1)-(2N_1+N_0)\psi\right] \quad . \tag{4.3.14}$$

Also, from (4.3.9), we obtain an expression for t_0:

$$t_0 = 27N_2^2N_1^2f\psi^3/(N_2-N_1)^2(N_1-N_0)(1-\psi^3)^2 \quad . \tag{4.3.15}$$

The domain of the parameter k is evidently unrestricted. However, examination of (4.3.9) and, with $r = 0$, (4.3.11) shows that only positive values need be considered. This can be seen by replacing k by -k in each of these formulas and noting that t_0 and q are unchanged. It follows that values obtained for these quantities for negative k only duplicate those obtained from positive values of the variable. Now, referring to (4.3.12), using only positive values of k, we can see that the range of ψ is over the interval (-1,1).

Now consider (4.3.9). Note that, in general, if k is complex, then t_0 is complex unless k is a pure imaginary. This is the second of the two cases we are to consider, which we will refer to as the *imaginary case*. For convenience, let us set

$$k = i \tan\theta \quad ,$$

where we assume that θ lies in the interval $(-\pi/2,\pi/2)$. Then

$$\left(\frac{k+1}{k-1}\right)^{1/3} = \left(\frac{i\ \sin\theta+\cos\theta}{i\ \sin\theta-\cos\theta}\right)^{1/3} = -e^{2i\theta/3} \quad .$$

It follows that

$$\omega^r\left(\frac{k+1}{k-1}\right)^{1/3} + \omega^{-r}\left(\frac{k-1}{k+1}\right)^{1/3} = -2\ \cos[2(\theta+\pi r)/3] \quad .$$

Substituting these expressions into (4.3.11) results in

$$q_r = 3N_1^2/\left\{2N_1+N_0+2(N_1-N_0)\ \cos[2(\theta+\pi r)/3]\right\} \quad ,$$

while (4.3.9) becomes

$$t_0 = -27N_1^2N_2^2f\ \sec^2\theta/4(N_2-N_1)^2(N_1-N_0) \quad .$$

However, everything becomes much simpler if we set

$$\phi = 2(\theta+\pi r)/3 \quad .$$

Then q_r and t_0 assume the same values if ϕ lies in the interval $(-\pi,\pi)$ for each of the three values of r. From the previous equations, now dropping the subscript r, we obtain

$$k = i\ \tan3\phi/2 \quad , \tag{4.3.16}$$

$$q = 3N_1^2/[2N_1+N_0+2(N_1-N_0)\ \cos\phi] \quad , \tag{4.3.17}$$

and

$$t_0 = \frac{-27N_2^2N_1^2f}{2(N_2-N_1)^2(N_1-N_0)(1+\cos3\phi)} \quad . \tag{4.3.18}$$

However, if ϕ is allowed to vary over the entire range $(-\pi,\pi)$, values of q and t_0 are duplicated. Therefore, with no loss of generality, we may restrict the domain of ϕ to the interval $(0,\pi)$.

4.4 Coma and Astigmatism

Either of the two pairs of equations, (4.3.14,15) or (4.3.17,18), when sub-stituted into the expression for spherical aberration (3.3.8), cause it to vanish. Therefore, from the first equation in (3.3.11), we may write

$$U_1 = t_0\gamma_1\alpha_1^2 + N_0 f\gamma_2\alpha_2^2 = 0 \quad ,$$

so that

$$\gamma_2 = -t_0\gamma_1\alpha_1^2/N_0 f\alpha_2^2 \quad . \tag{4.4.1}$$

Making use of this expression, we find that the expression for coma, (3.3.9), becomes

$$\delta = u_0^3\bar{y}_0 U_2 \quad , \tag{4.4.2}$$

and the expression for astigmatism, (3.3.10), turns into

$$c = u_0^2\bar{y}_0(2\bar{u}_0 U_2 + \bar{y}_0 U_3) \quad . \tag{4.4.3}$$

As the result of a most fortunate factorization, (4.4.3) may be written in the form

$$c = u_0^2\bar{y}_0 t_0\gamma_1(\alpha_2\beta_1 - \alpha_1\beta_2)[2\bar{u}_0\alpha_1\alpha_2 + \bar{y}_0(\alpha_2\beta_1 + \alpha_1\beta_2)]/\alpha_2^2 \quad .$$

By setting this expression for astigmatism equal to zero, we get either

$$\alpha_2\beta_1 - \alpha_1\beta_2 = 0 \tag{4.4.4}$$

or

$$2\bar{u}_0\alpha_1\alpha_2 + \bar{y}_0(\alpha_2\beta_1 + \alpha_1\beta_2) = 0 \quad . \tag{4.4.5}$$

Note that (4.4.4) does not depend on the chief-ray variables $\bar{u}_0$ and $\bar{y}_0$, where-as (4.4.5) does.

We consider the implications of (4.4.4) first. Substituting from (3.3.4), we find that it reduces to

$$N_2 f(N_0 - q) - t_0(N_1 - q) = 0 \quad . \tag{4.4.6}$$

It turns out that this is exactly the case in which the module is concentric. We can see this by setting

$$1/c_1 = t_1 + 1/c_2 \quad ,$$

and applying (3.1.8,9). Whenever a lens is concentric, both astigmatism and coma are zero. We will see shortly that lateral chromatic aberration also vanishes.

Equation (4.4.6) may be satisfied by one or more values of ψ or ϕ. These zeros will be a set of critical values, to be treated in a subsequent chapter.

Next, we take up the case of (4.4.5). This we can write in the form

$$\bar{y}_0/\bar{u}_0 = -2\alpha_1\alpha_2/(\alpha_2\beta_1 + \alpha_1\beta_2) \quad .$$

When the ratio of the height of the chief ray at the main focus to its slope satisfies this relation, then third-order astigmatism vanishes. Now, referring to a cross-over equation, such as (3.1.2), we see that the point where the ray crosses the axis is given by

$$\bar{t} = 2\alpha_1\alpha_2/(\alpha_2\beta_1 + \alpha_1\beta_2) \quad ,$$

which is exactly the location of a pupil plane. The distance $\bar{t}$ is, of course, measured from the main focus.

If either of these expressions holds, that is, if either (4.4.4) or (4.4.5) is satisfied, then astigmatism is zero and, referring to (4.4.3), we may write

$$2\bar{u}_0 U_2 + \bar{y}_0 U_3 = 0 \quad , \tag{4.4.7}$$

or

$$\bar{t} = 2U_2/U_3 \quad . \tag{4.4.8}$$

We designate this location as the *front-pupil plane*. Substituting from (3.3.4), we obtain

$$\bar{t} = \frac{2N_2 t_0 f q(N_1 - q)}{N_1 t_0(N_1 - q) - N_2 f(N_0 - q)(N_1 - 2q)} \quad . \tag{4.4.9}$$

The chief ray, by definition, passes through the axial point of the pupil plane. Third-order astigmatism therefore vanishes for every choice of chief ray, once the location of the pupil plane is determined by (4.4.7).

For the location of the *back-pupil plane*, we need only substitute the expression for $\bar{t}$ in (4.4.9) into (3.2.5), the expression for finding a conjugate plane. The result is

$$\bar{t}' = -N_0 f \; \frac{t_0(N_1 - 2N_2)(N_1 - q) + N_2 N_1 f(N_0 - q)}{2t_0 q(N_1 - q)} \; . \tag{4.4.10}$$

We will also need a plane conjugate to these pupil planes, between the two surfaces. Again we invoke (3.1.2) and calculate

$$\tilde{t} = -\bar{y}_1/\bar{u}_1 = -N_1 t_0(t_0 - \bar{t})/[(N_0 - q)\bar{t} + t_0 q]$$

in order to get

$$\tilde{t} = \frac{t_0}{q} \; \frac{f N_2 [N_1 q + N_0(N_1 - 2q)] + N_1 t_0(N_1 - q)}{N_2 f(N_0 - q) + t_0(N_1 - q)} \; . \tag{4.4.11}$$

The derivation of (4.4.11) is difficult, in the absence of some sort of guidance. Begin by defining the cross-over point for the chief ray in the medium between the two surfaces,

$$t = -\bar{y}_1/\bar{u}_1 \quad ,$$

from (3.1.2). From (3.1.1), we have

$$\bar{y}_1 = \bar{y}_0 + t_0 \bar{u}_0 \quad ,$$

$$\bar{u}_1 = -(1 - \mu_1)c_1 \bar{y}_0 + [\mu_1 - (1 - \mu_1)c_1 t_0]\bar{u}_0 \quad .$$

But, from the argument that leads up to (4.4.8),

$$\bar{y}_0 = -\bar{t}\bar{u}_0 \quad ,$$

so that

$$\bar{y}_1 = (t_0 - \bar{t})\bar{u}_0 \quad .$$

58

The expression for $\bar{u}_1$ then becomes

$$\bar{u}_1 = [(N_0-q)\bar{t}+t_0q]/N_1t_0$$

so that

$$\tilde{t} = -N_1t_0(t_0-\bar{t})/[(N_0-q)\bar{t}+t_0q] \quad .$$

Making the appropriate substitution from (4.4.8) yields (4.4.11). We need
the last equation for the case of a module whose second medium is air. A real
stop can then be located at this point.

This completes the definition of the module that consists of two refracting spherical surfaces that separate three media, a main focus for which
third-order spherical aberration is zero, and a pair of pupil planes for which
third-order astigmatism is zero. A plane conjugate to the pupil planes in the
medium between the two surfaces is included.

Figure 4.1 shows a schematic diagram of the module in its forward orientation; Fig.4.2 shows its y,$\bar{y}$ diagram. Figures 4.3 and 4.4 show graphs of

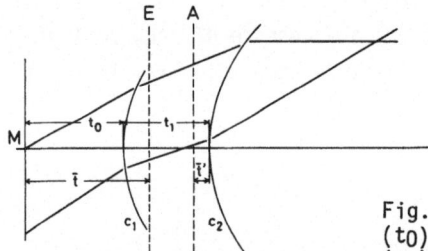

Fig. 4.1. The module. (M) Main focus;
(t_0) object distance; (t_1) axial separation;
(c_2) curvature, 2nd surface;
(t) distance, main focus to front pupil; (t') distance, 2nd surface to back
pupil; (c_1) curvature, 1st surface

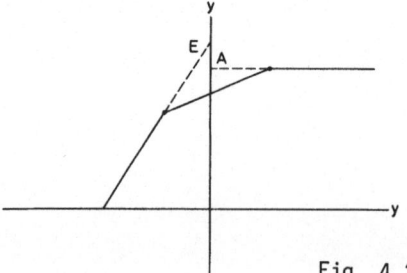

Fig. 4.2. The module. y-y diagram. (E) Entrance
pupil; (A) Exit pupil

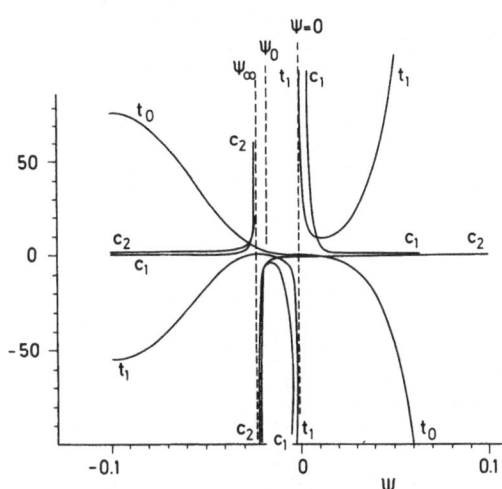

Fig. 4.3. The optical parameters vs ψ, real case. t_0 and t_1 are in units of focal length, c_1 and c_2 in units of inverse focal length. Both are drawn to the same scale

Fig. 4.4. The optical parameters vs ϕ, imaginary case. t_0 and t_1 are in units of focal length, c_1 and c_2 in units of inverse focal length. Both are drawn to the same scale

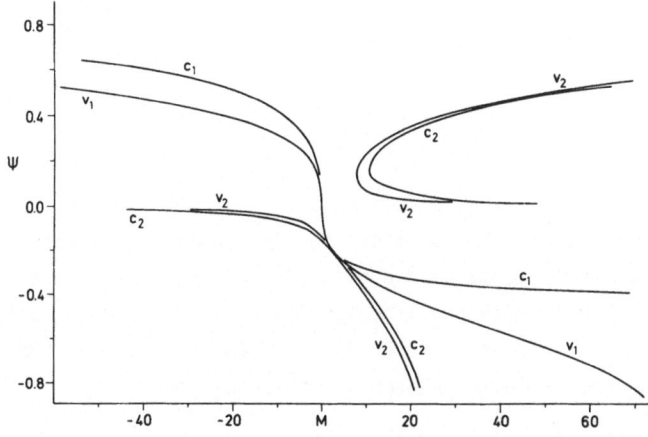

Fig. 4.5. Graphical construction, real case. $N_0 = 1.8$, $N_1 = 1.0$, $N_2 = 1.5$. Construct a horizontal line parallel to the abscissa at a height equal to the desired value of ψ. Where it intersects curves V_1 and V_2 marks the locations of the vertices of the two spherical surfaces. Where it intersects curves C_1 and C_2 marks the locations of the two centers. M marks the location of the main focus. t_1 should be positive. If it is negative, then flip the figure around the line marked M to change all signs

60

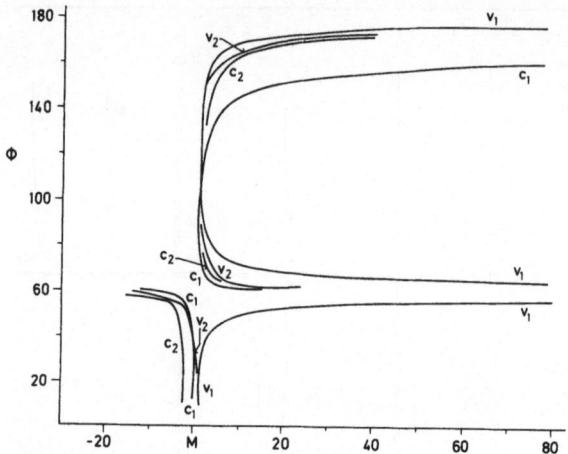

Fig. 4.6. Graphical construction, imaginary case. $N_0 = 1$, $N_1 = -1$, $N_2 = 1$. A reflecting module. Proceed as in Fig.4.5. Here t_1 should be negative. If t_1 is positive, then flip the obtained figure around the line marked M

the various optical parameters for a real-case and an imaginary-case module. Figures 4.5 and 4.6 show a graphic method for laying out a module when the three refractive indices are known.

4.5 The Pupil Errors

We are now in a position to render the equations for the pupil aberrations: pupil spherical aberration, pupil coma, and pupil astigmatism. We will use (3.3.11) and (3.4.1-6). We begin with pupil astigmatism, as provided in (3.4.1) and the associated anciliary formulas in (3.3.11). From (4.4.1), $U_1 = 0$, and from (4.4.7), $2\bar{u}_0 U_2 + \bar{y}_0 U_3 = 0$, so that the expression reduces to

$$\bar{c} = u_0^2 \bar{y}_0 [\bar{u}_0 (W_1 + W_2) + \bar{y}_0 W_3] \quad . \tag{4.5.1}$$

Pupil coma is next. We use (3.4.2-4) and obtain

$$\bar{\delta} = u_0 \bar{y}_0 [\bar{u}_0^2 (U_2 + W_1 + W_2) + \bar{u}_0 \bar{y}_0 (W_3 + V_1 + V_2) + \bar{y}_0^2 V_3] \quad . \tag{4.5.2}$$

By use of (4.5.1) and (4.4.2), this becomes

$$\bar{\delta} = \delta(\bar{u}_0/u_0)^2 + \bar{c}(\bar{u}_0/u_0) + u_0\bar{y}_0^2[\bar{u}_0(V_1+V_2)+\bar{y}_0V_3] \quad . \tag{4.5.3}$$

Next comes spherical aberration. Here, we make use of (3.4.5,6) as well as (4.4.1,7)

$$\bar{b} = \bar{y}_0\left[\bar{u}_0^3(W_1+W_2)+\bar{u}_0^2\bar{y}_0(2V_1+2V_2+W_3)+\bar{u}_0\bar{y}_0^2(2V_3+V_4+V_5)+\bar{y}_0^3V_6\right] \quad . \tag{4.5.4}$$

As in the case of pupil coma, this can also be reduced by using (4.5.1-3) as well as (4.4.2), which yields

$$\bar{b} = -2\delta(\bar{u}_0/u_0)^3 - \bar{c}(\bar{u}_0/u_0)^2 + 2\bar{\delta}(\bar{u}_0/u_0) + \bar{y}_0^3[\bar{u}_0(V_4+V_5)+\bar{y}_0V_6] \quad . \tag{4.5.5}$$

4.6 Field and Chromatic Aberrations

We next turn to the aberrations associated with the image of an extended object. We have already derived a formula for the Petzval contribution in (3.5.1) in the previous chapter which we repeat here for the convenience of the reader,

$$p = [N_2f(N_0-1)+t_0q]/N_2N_1N_0ft_0 \quad . \tag{4.6.1}$$

In (3.5.2), we obtained an expression for image distortion,

$$e = \bar{\delta} - h\left(\bar{y}_0^2-N_2^2f^2\bar{u}_0^2\right)/N_2^2f^2 \quad , \tag{4.6.2}$$

and in (3.5.3) one for pupil distortion,

$$\bar{e} = \delta - hu_0^2 \quad . \tag{4.6.3}$$

To obtain the proper expressions for these aberrations for modules, we need only substitute in the appropriate expressions for pupil coma (4.5.2), image coma (4.4.2), and the relationship between $\bar{y}_0$ and $\bar{u}_0$ given in (4.4.7).

Similarly, the chromatic aberrations found in (3.6.1-4) stand as given. Longitudinal and transverse color are in the form

$$\ell = u_0^2\hat{\ell} \quad , \tag{4.6.4}$$

$$\hat{t} = u_0(\bar{u}_0\hat{L}+\bar{y}_0\hat{T}) \quad , \tag{4.6.5}$$

where

$$\hat{L} = \alpha_1 t_0(D_1/N_1-D_0/N_0) + N_0 f\alpha_2(D_2/N_2-D_1/N_1) \tag{4.6.6}$$

and

$$\hat{T} = t_0\beta_1(D_1/N_1-D_0/N_0) + \beta_2 N_0 f(D_2/N_2-D_1/N_1) \quad . \tag{4.6.7}$$

Here we use the notation introduced in (3.3.4). Of course, in the application of these formulas we need to use (4.4.7).

5. Critical Values

The expression for q, whether considered as a function of ψ or of ϕ is particularly wild. The expression for t_0 in terms of these same variables is only slightly tamer. In this chapter, we will seek values of these independent variables that result in zeros or poles for q and t_0, as well as the optical parameters associated with the module. This investigation will do more than provide a map of points at which a module undergoes a profound morphological change. These points are boundaries for regions in which the module is reasonably well behaved. A knowledge of these regions is particularly important in the coupling problem. We will begin by defining some properties of the module that effect its coupling. In what follows, the work involving the imaginary case is largely that of MERCADO [5.1].

5.1 Categories

We define two categories that determine the easy-way coupling properties of a module. This is done in Table 5.1. These categories are concerned with whether the distance from the second surface to the second pupil is positive or negative. Recall that the easy-way couple occurs when the second pupils of a pair of modules are made to coincide (Fig.5.1). The distance between the two second surfaces, of course, must be positive for successful coupling. This can certainly happen when both modules are in catergory X, as shown in Fig.5.1; indeed, both a necessary and a sufficient condition for successful coupling is satisfied. Moreover, the pupil is real, not virtual, and its location is a candidate for an aperture stop.

Table 5.1. Easy-way coupling categories

Condition	$\bar{t}' > 0$	$\bar{t}' < 0$
Category	X	Y

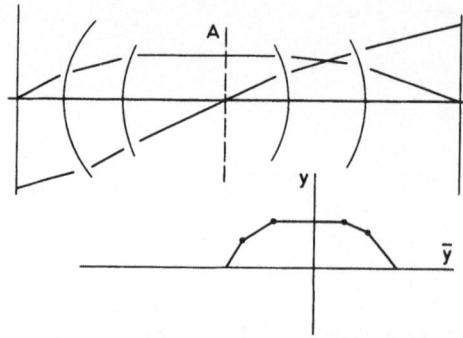

Fig. 5.1. Easy-way coupling. XX
(A) denotes the common pupil plane

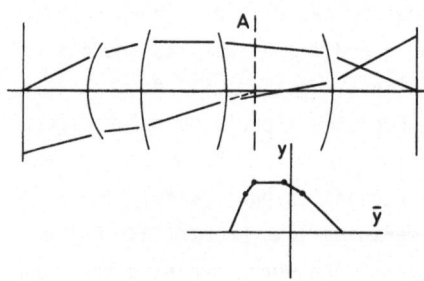

Fig. 5.2. Easy-way coupling. XY (A) denotes the common pupil plane

If one of the modules is in category X and the other is in category Y,
then a necessary (but not a sufficient) condition for successful coupling is
satisfied. A sufficient condition requires that the magnitude of $\bar{t}'$ in the X
module exceeds the absolute value of $\bar{t}'$ in the Y module. In this case, the
pupil is virtual and it is not possible to place an aperture stop at its location. This is shown in Fig.5.2. If both modules are in the Y category, then
coupling is indeed impossible and neither a necessary not a sufficient condition is satisfied.

We can do the same kind of analysis for the hard-way couple. The appropriate categories are defined in Table 5.2. These four categories are concerned
with the main focus of the module, where hard-way coupling occurs.

Table 5.2. Hard-way coupling categories

	$\bar{t} > 0$	$\bar{t} < 0$
$t_0 > 0$	A	B
$t_0 < 0$	C	D

Figures 5.3-7 show several combinations of the easy-way and hard-way coupling categories. Figure 5.3 gives a sketch of a module in category AX along

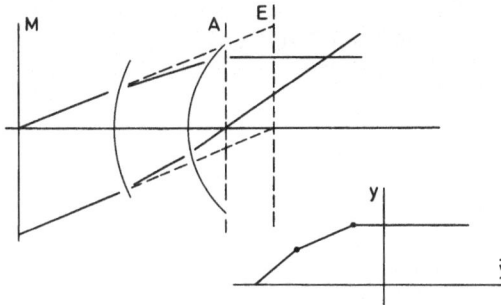

Fig.5.3. Category AY

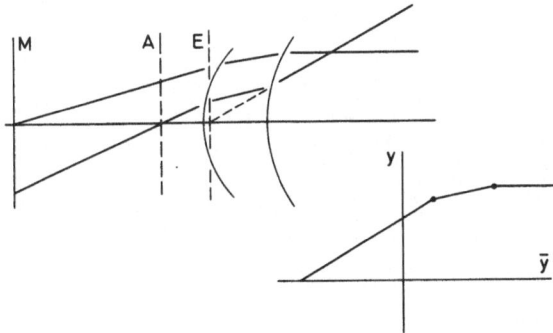

Fig. 5.4. Category AX

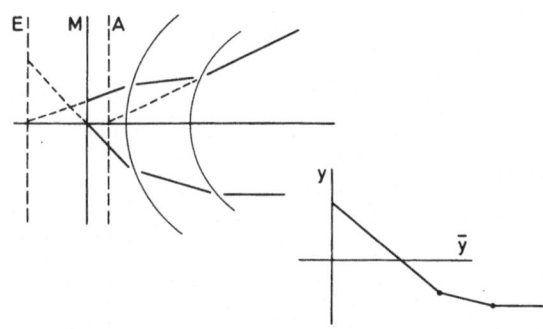

Fig. 5.5. Category BY.
Note from the y-ȳ diagram
that category BX is not
possible

with its y-ȳ diagram. Figure 5.4 shows category AY, and Fig.5.5 category BY.
Note that by means of the y-ȳ diagram it is a simple matter to see that cate-
gory BX cannot exist. Figure 5.6 illustrates category CX, and Fig.5.7 category
DY. In both cases, we can see that categories CY and DX can be constructed.

A necessary and sufficient condition for coupling is satisfied if one of
the modules is in category A and the other in category B, as shown in Fig.5.8.

66

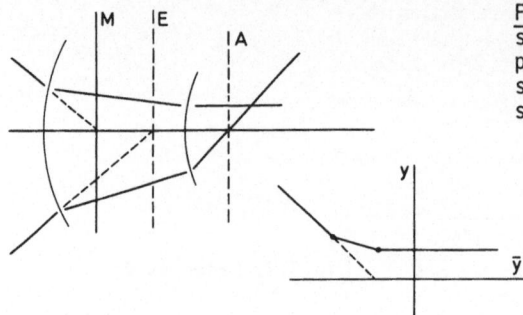

Fig. 5.6. Category CX. Note that by shifting the point representing the point that represents the second surface, category CY can be constructed

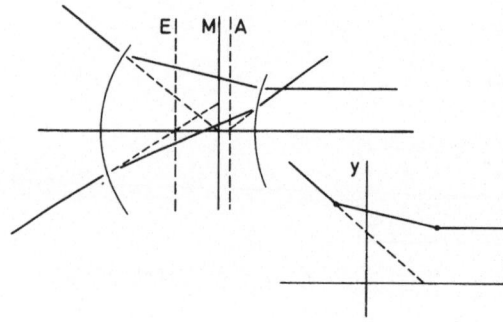

Fig. 5.7. Category DY. By shifting the point that represents the second surface, category DX can be constructed

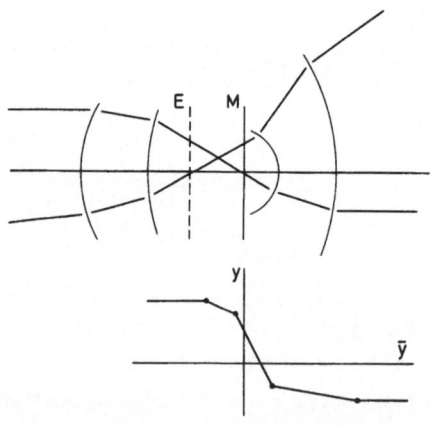

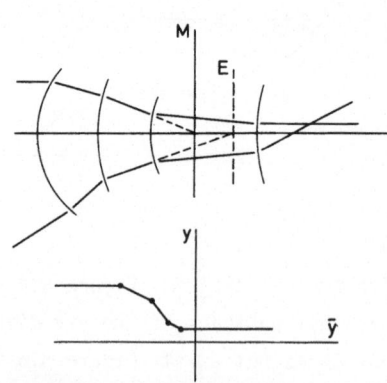

Fig. 5.8. Hard-way coupling, AB. (M) and (E) denote common main foci and pupil planes

Fig. 5.9. Hard-way coupling, AD. (M) and (E) denote common main foci and pupil planes

A necessary (but not sufficient) condition is satisfied if the categories are A and D or B and C, shown in Figs.5.9,10. All other combinations are impossible.

Note that in both the A,B coupling and the A,D coupling the possibility of a real pupil exists. However, in the case of the C,B coupling, the pupil must be virtual.

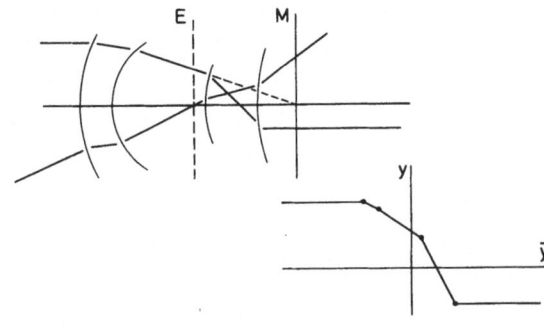

Fig. 5.10. Hard-way coupling, BC. (M) and (E) denote common main foci and pupil planes

5.2 The Method

In obtaining the critical values that we seek, we must remember that two distinct cases, the real case and the imaginary case, must be accounted for. In general, each class of singularity will call for a certain equation, usually involving t_0 and q, to be solved for the shape parameter. We would expect to have to use (4.3.14) and (4.3.15), the equations for q and t_0, respectively, in the real case and then (4.3.17) and (4.3.18) for the imaginary case. In principle, we should be able to use (4.3.9) and (4.3.11) in the general case, and solve for k. Then, if k is real, we would use (4.3.12) to obtain the appropriate value of ψ, whereas if k is pure imaginary we would obtain a value of ϕ by application of (4.3.16).

However, there is a much simpler approach, which enables us to slay both birds with the same boulder. Note, first of all, that (4.3.16),

$$k = i \tan 3\phi/2 \quad ,$$

and (4.3.12),

$$\psi = (k-1)/(k+1) \quad ,$$

enable us to write ψ as a function of ϕ, so that

$$\psi = -e^{-i\phi} = -\cos\phi + i\,\sin\phi \quad . \tag{5.2.1}$$

Now we see that we need only set up the equations for the critical values, using the real-case equations alone; that is, using the expressions for q and t_0 given by (4.3.14) and (4.3.15), respectively. Then, any real solution gives us a real-case value of ψ. A complex solution, on the other hand, gives us an imaginary-case solution for ϕ by means of (5.2.1), provided that the complex solution is unimodular, that $|\psi| = 1$.

5.3 The Simple Singularities

We refer to the zeros and poles of t_0, q, and c_2 as the simple singularities of the module. Not only are they simple, they are the least important. Their signal property is that they depend on only N_0 and N_1 and not on N_2. The only thing that can be said in their favor is that they are easy to do and provide a convenient model for the subsequent, more difficult cases.

We begin with t_0, given by (3.3.15):

$$t_0 = 27N_2^2 N_1^2 f\psi^3 / (N_2-N_1)^2 (N_1-N_0)(1-\psi^3)^2 \quad .$$

No great insight is required to see that a zero occurs at $\psi = 0$ and that poles occur at $\psi = 1$ and at the complex cube roots of unity, $\psi = \exp(\pm 2\pi i/3)$. Making use of (5.2.1), we see that the latter correspond to $\phi = \pm 2\pi/3$. These results are summarized in Table 5.3.

Table 5.3. Critical values from t_0

	Real case (ψ)	Imaginary case (ϕ)
Zeros	0	—
Poles	1	$2\pi/3$

We treat q and c_2 as one, since they are connected by (3.1.8),

$$c_2 = q/N_0(N_2-N_1)f .$$

We use (4.3.14) for q:

$$q = -3N_1^2\psi / [(N_1-N_0)(\psi^2+1)-(2N_1+N_0)\psi] .$$

It is not difficult to see that a zero for q occurs at $\psi = 0$ and that poles appear as solutions of the quadratic polynomial,

$$(N_1-N_0)\psi^2 - (2N_1+N_0)\psi + (N_1-N_0) = 0 .$$

Here we use (4.2.2) and obtain the solutions

$$\psi_\pm = \frac{(4N_1-N_0)^{\frac{1}{2}} \mp (3N_0)^{\frac{1}{2}}}{(4N_1-N_0)^{\frac{1}{2}} \pm (3N_0)^{\frac{1}{2}}} .$$

If $4N_1-N_0 > 0$, then these two roots lie in the real case. Because the absolute value of ψ must be less than unity, we may write the solution as

$$\psi_\infty = \frac{(4N_1-N_0)^{\frac{1}{2}} - (3N_0)^{\frac{1}{2}}}{(4N_1-N_0)^{\frac{1}{2}} + (3N_0)^{\frac{1}{2}}} . \tag{5.3.1}$$

Not only is this expression real when $4N_1-N_0 > 0$, but it is positive as long as $N_1-N_0 > 0$.

In the unlikely event that $4N_1-N_0 < 0$, then the root becomes complex:

$$\psi = \frac{i(N_0-4N_1)^{\frac{1}{2}} - (3N_0)^{\frac{1}{2}}}{i(N_0-4N_1)^{\frac{1}{2}} + (3N_0)^{\frac{1}{2}}} . \tag{5.3.2}$$

The easiest approach to this is to write (4.2.1) in the form

$$\psi = -e^{-i\phi} = -\frac{\exp(-i\phi/2)}{\exp(i\phi/2)} = -\frac{\cos(\phi/2) - i \sin(\phi/2)}{\cos(\phi/2) + i \sin(\phi/2)} .$$

By comparing corresponding terms in this expression and that in (5.3.2), we eventually obtain

$$\phi_\infty = \text{arc tan} \frac{(3N_0)^{\frac{1}{2}}(N_0-4N_1)^{\frac{1}{2}}}{N_0 + 2N_1} . \tag{5.3.3}$$

These results are summarized in Table 5.4.

Table 5.4. Critical values from q

	Real case (ψ)	Imaginary case (ϕ)
Zeros	0	—
Poles	$4N_1 - N_0 > 0$	$4N_1 - N_0 < 0$
	ψ_∞	ϕ_∞

5.4 Singularities for the Optical Parameters

The optical parameters treated in this section are t_1 and c_1, which are given in (3.1.9-10). It is most convenient to treat c_1 first. From (3.1.10), we have

$$c_1 = (N_0 - q) / (1 - \mu_1) N_1 t_0 \ .$$

The zeros of c_1 occur at the poles of t_0, given in Table 5.3, and when $q = N_0$. The poles of c_1, quite aptly, lie on the zeros of t_0, which also can be found in Table 5.3, and at the poles of q, ψ_∞, or ϕ_∞ given in Table 5.4. The one equation that we must deal with is obtained, therefore, from $q = N_0$, which, using (4.3.14), leads to

$$-3N_1^2 \psi = N_0 [(N_1 - N_0)(\psi^2 + 1) - (2N_1 + N_0)\psi] \ ,$$

which reduces quickly to

$$N_0 \psi^2 + (3N_1 + N_0)\psi + N_0 = 0 \ ,$$

a quadratic in ψ. Referring once again to (4.2.2), we obtain

$$\psi_\pm = - \frac{[3(N_1 + N_0)]^{\frac{1}{2}} \mp (3N_1 - N_0)^{\frac{1}{2}}}{[3(N_1 + N_0)]^{\frac{1}{2}} \pm (3N_1 - N_0)^{\frac{1}{2}}} \ . \qquad (5.4.1)$$

Note that if the first surface is a mirror, so that $N_1 = -N_0$, then $\psi = 1$. Also recall that $|\psi| < 1$, so that the sign ambiguity can be suppressed; we therefore drop $\pm$ in (5.4.1) and designate the quantity as ψ_0.

The expression for ψ_0 is real whenever $3N_1 - N_0 > 0$. We exclude as unrealistic the possibility that $N_1 + N_0 < 0$.

When $3N_1 - N_0 < 0$, we are faced with the imaginary case. Using the same trick that served us so well in Sect.5.3, in conjunction with (5.2.1), we obtain

$$\phi_0 = \text{arc tan} \frac{[3(N_1+N_0)(N_0-3N_1)]^{\frac{1}{2}}}{2(3N_1+N_0)} . \tag{5.4.2}$$

These results are summarized in Table 5.5.

Table 5.5. Critical values associated with c_1

	Real case (ψ)	Imaginary case (ϕ)
Zeros	1	$2\pi/3$
	—	$-2\pi/3$
	—	2π
	$3N_1 - N_0 > 0$	$3N_1 - N_0 < 0$
	ψ_0	ϕ_0
Poles	0	—
	$4N_1 - N_0 > 0$	$4N_1 - N_0 < 0$
	ψ_∞	ϕ_∞

We now come to the first really difficult calculation for critical values. These are associated with the thickness, t_1, given by (3.1.9),

$$t_1 \doteq N_1(N_0f-t_0)/q .$$

The zeros of t_1 occur at the poles of q, ψ_∞, or ϕ_∞, which are found in Table 5.4, and when $N_0f - t_0 = 0$. Its poles lie on the poles of t_0 and on the zeros of q. Using (4.3.15), we set $t_0 = N_0f$ and obtain

$$27N_2^2N_1^2\psi^3 = N_0(N_2-N_1)^2(N_1-N_0)(1-\psi^3)^2 ,$$

which is a quadratic in ψ^3 in the form

$$A\psi^6 + B\psi^3 + C = 0 ,$$

where $A = C = N_0(N_2-N_1)^2(N_1-N_0)$ and $B = -2A - 27N_2^2N_1^2$. Now, referring once again to (4.2.2), we have

$$\psi^{*3} = \frac{[1+4N_0(N_2-N_1)^2(N_1-N_0)/27N_2^2N_1^2]^{\frac{1}{2}} - 1}{[1+4N_0(N_2-N_1)^2(N_1-N_0)/27N_2^2N_1^2]^{\frac{1}{2}} + 1} \quad . \tag{5.4.3}$$

Whenever this expression is real, it lies in the interval $(-1,1)$. It is real whenever

$$27N_2^2N_1^2 + 4N_0(N_2-N_1)^2(N_1-N_0) > 0 \quad . \tag{5.4.4}$$

This condition is always satisfied when $N_0(N_1-N_0) > 0$. However when $N_0(N_1-N_0) < 0$, the expression is real for only certain values of N_2, N_1, and N_0. To see what these values are, write the left member of (5.4.4) as a quadratic in N_2. This factors, resulting in

$$[N_2(3\sqrt{3}N_1+2P)-2N_1P][N_2(3\sqrt{3}N_1-2P)+2N_1P] > 0 \quad ,$$

where

$$p^2 = N_0(N_0-N_1) \quad . \tag{5.4.5}$$

For this inequality to hold, both factors must be either positive or negative. Let

$$\begin{aligned} A &= N_2(3\sqrt{3}N_1+2P) - 2N_1P \quad , \\ B &= N_2(3\sqrt{3}N_1-2P) + 2N_1P \quad . \end{aligned} \tag{5.4.6}$$

It follows that either $A > 0$ and $B > 0$ or $A < 0$ and $B < 0$ if ψ^3, given in (5.4.3), is real. For ψ^3 to be complex, it must be that

$$-2N_1p / (3\sqrt{3}N_1-2p) < N_2 < 2N_1p / (3\sqrt{3}N_1+2p) \quad . \tag{5.4.7}$$

In this case, we may write ψ^3 as

$$\psi^3 = -\frac{3\sqrt{3}N_2N_1 - i[4p^2(N_2-N_1)^2-27N_2^2N_1^2]^{\frac{1}{2}}}{3\sqrt{3}N_2N_1 + i[4p^2(N_2-N_1)^2-27N_2^2N_1^2]^{\frac{1}{2}}} = \exp\left\{-2i[4p^2(N_2-N_1)^2/27N_2^2N_1^2-1]^{\frac{1}{2}}\right\} \quad .$$

Referring to (5.2.1), we see that the critical value of ϕ for the imaginary case is

$$\phi = \arctan\left\{(2/3)[4p^2(N_2-N_1)2/27N_2^2N_1^2-1]^{\frac{1}{2}}\right\} . \tag{5.4.8}$$

These results are summarized in Table 5.6.

Table 5.6. Critical values associated with t_1

	Real case (ψ)	Imaginary case (ϕ)
Zeros	$4N_1 - N_0 > 0$	$4N_1 - N_0 < 0$
	ψ_∞	ϕ_∞
	$N_0(N_1-N_0) > 0$ or	$N_0(N_1-N_0) < 0$ and
	$AB > 0$	$AB > 0$
	ψ^*	ϕ^*
Poles	0	—
	1	$2\pi/3$

$$A = N_2(3\sqrt{3}N_1-2P) - 2N_1P$$
$$B = N_2(3\sqrt{3}N_1-2P) + 2N_1P$$
$$p^2 = N_0(N_0-N_1)$$

5.5 Pupil Singularities

These occur when $\bar{t}$ or $\bar{t}'$ have poles or zeros and clearly differentiate several distinct categories. Let us begin with $\bar{t}$ and (4.4.9):

$$\bar{t} = \frac{2N_2t_0fq(N_1-q)}{N_1t_0(N_1-q) - N_2f(N_0-q)(N_1-2q)} . \tag{5.5.1}$$

Obviously, the zeros of $\bar{t}$ are the zeros of t_0 and q; these can be found in Tables 5.3,4. We might also guess that additional zeros for $\bar{t}$ would correspond to the zeros of $N_1 - q = 0$, but, as we shall see, this is not the case. Note that

$$N_1 - q = N_1(N_1-N_0)(\psi^2+\psi+1)/U , \tag{5.5.2}$$

where

$$U = (N_1 - N_0)(\psi^2 + 1) - (2N_1 + N_0)\psi \quad . \tag{5.5.3}$$

Here we have made use of (4.3.14).

Poles for $\bar{t}$ occur as zeros of the denominator of (5.5.1). Substituting for t_0 and q, and using (4.3.14-15), we obtain, after clearing fractions and making several cancellations

$$N_2(N_2-N_1)^2(N_1-N_0)(1-\psi^3)^2[N_0(\psi^2+1)+(3N_1+N_0)\psi][(N_1-N_0)(\psi^2+1)+(4N_1-N_0)\psi]$$

$$- 27N_2^2 N_1^3 \psi^3 (\psi^2+\psi+1)[(N_1-N_0)(\psi^2+1)-(2N_1+N_0)\psi] = 0 \quad .$$

Note, first of all, that $1 - \psi^3 = (1-\psi)(\psi^2+\psi+1)$ so that the expression $\psi^2 + \psi + 1$ appears as a factor of the entire equation. However, this does not lead to a singularity because it cancels exactly the factor $N_1 - q$ that appears in the numerator and is given in (5.5.2). Equation (5.5.3), thus, can be reduced to

$$(N_2-N_1)^2(N_1-N_0)(\psi^4-\psi^3-\psi+1)[N_0(\psi^2+1)+(3N_1+N_0)\psi][(N_1-N_0)(\psi^2+1)+(4N_1-N_0)\psi]$$

$$- 27N_2 N_1^3 \psi^3 [(N_1-N_0)(\psi^2+1)-(2N_1+N_0)\psi] = 0 \quad . \tag{5.5.4}$$

This is obviously a polynomial of degree 8 in ψ. It is less obvious that it is a reciprocal polynomial in ψ and, as such, its degree may be quickly and easily reduced to degree 4. A *reciprocal polynomial* is one which remains unchanged when the variable is replaced by its reciprocal.

To illustrate this, consider any reciprocal polynomial, say, one of degree 4:

$$ax^4 + bx^3 + cx^2 + bx + a = 0 \quad .$$

By dividing by x^2 and rearranging terms, we can get

$$a(x^2+1/x^2) + b(x+1/x) + c = 0 \quad .$$

Now let $z = x + 1/x$. Then $x^2 + 1/x^2 = z^2 - 2$. Substituting yields the quadratic

$$az^2 + bz + c - 2a = 0 \quad ,$$

which is easy to solve. Each of the two roots, z, is now substituted into the quadratic

$$x^2 - zx + 1 = 0 \quad , \tag{5.5.5}$$

whose roots, in turn, are exactly the four roots of the original quartic.

We follow this procedure in (5.5.4), first dividing by ψ^4 and then substituting $\psi + 1/\psi = z$ and obtain

$$(N_2-N_1)^2(N_1-N_0)(z^2-z-2)(N_0z+3N_1+N_0)[(N_1-N_0)z+4N_1-N_0]$$

$$- 27N_2N_1^2[(N_1-N_0)z-(2N_1+N_0)] = 0 \quad .$$

After expanding this expression and collecting terms, we obtain

$$az^4 + bz^3 + cz^2 + dz + e = 0 \quad ,$$

where

$$a = N_0(N_2-N_1)^2(N_1-N_0)^2 \quad ,$$

$$b = (N_2-N_1)^2(N_1-N_0)(3N_1^2+N_1N_0-N_0^2) \quad ,$$

$$c = 3(N_2-N_1)^2(N_1-N_0)(3N_1^2-N_1N_0+N_0^2) \quad ,$$

$$d = -(N_2-N_1)^2(N_1-N_0)(18N_1^2+5N_1N_0-5N_0^2) - 27N_2N_1^3(N_1-N_0) \quad ,$$

$$e = -2(N_2-N_1)^2(N_1-N_0)(12N_1^2+N_1N_0-N_0^2) + 27N_2N_1^3(2N_1+N_0) \quad . \tag{5.5.6}$$

It would be nice to say that we can solve the equation in a formal way; however, it appears to be beyond us and we must be content to obtain the roots of (5.5.6) on a computer. A discussion of the quartic is included later in this chapter, for the benefit of anyone who cares to attempt a formal treatment. The roots obtained are substituted into the quadratic, (5.5.5), whose roots provide the eight roots of (5.5.4). The real roots correspond to the real-case singularities, whereas the unimodular complex roots yield those for the imaginary case. Let us denote this class of critical values by $\bar{\psi}$ and $\bar{\phi}$.

We next consider the singularities associated with $\bar{t}'$, as given by (4.4.10),

$$\bar{t}' = -N_0f \frac{t_0(N_1-2N_2)(N_1-q) + N_2N_1f(N_0-q)}{2t_0q(N_1-q)} \quad . \tag{5.5.7}$$

We repeat the same ritual used in the other cases; the poles are the zeros of t_0, q, and $\psi^2 + \psi + 1 = 0$, whereas the zeros correspond to the poles of t_0, q, and

$$t_0(N_1-2N_2)(N_1-q) + N_2N_1f(N_0-q) = 0 \quad , \tag{5.5.8}$$

the numerator of (5.5.7).

Again making use of (3.14-15), we obtain, this time, a sextic in ψ,

$$(N_2-N_1)^2(N_1-N_0)(\psi^4-\psi^3-\psi+1)[N_0(\psi^2+1)+(3N_1+N_0)\psi]$$

$$+ 27N_2N_1^2(N_1-2N_2)\psi^3 = 0 \quad . \tag{5.5.9}$$

This also is reciprocal and reduces to the cubic

$$az^3 + bz^2 + cz + d = 0 \quad ,$$

where

$$a = N_0(N_2-N_1)^2(N_1-N_0) \quad ,$$

$$b = 3N_1(N_2-N_1)^1(N_1-N_0) \quad ,$$

$$c = -3(N_2-N_1)^2(N_1^2-N_0^2) \quad ,$$

$$d = -2(N_2-N_1)^2(N_1-N_0)(3N_1+N_0) + 27N_2N_1^2(N_1-2N_2) \quad . \tag{5.5.10}$$

The three roots of (5.5.10) are substituted for z in (5.5.5), a quadratic whose roots yield the six solutions of (5.5.9). The real solutions are the singularities of the real case, whereas the unimodular complex roots provide the imaginary-case singularities. We denote these critical values a $\bar{\psi}'$ and $\bar{\phi}'$.

5.6 Stop Singularities

These are the singularities associated with $\tilde{t}$; calculation of them is not quite routine. As in the cases of the other parameters considered, we are interested in the zeros and poles of $\tilde{t}$, but also in the case in which $\tilde{t} = t_1$. The region $0 < \tilde{t} < t_1$ is where the pupil in the space between the two surfaces of the module is real. This condition is of paramount importance when the medium is air and the designer wishes to locate a stop there. Refer to (4.4.11):

$$\tilde{t} = \frac{t_0}{q} \frac{fN_2[N_1q+N_0(N_1-2q)] - N_1t_0(N_1-q)}{N_2f(N_0-q) + t_0(N_1-q)} \quad . \tag{5.6.1}$$

The zeros of the factor t_0/q are the zeros of t_0 and the poles of q; they have been treated in Sect.5.3. The additional zeros are obtained by setting the numerator of the expression in (5.6.1) equal to zero

$$fN_2[N_1q+N_0(N_1-2q)] - N_1t_0(N_1-q) = 0 \quad . \tag{5.6.2}$$

Proceeding as before, we substitute from (4.3.14-15) and obtain

$$(N_2-N_1)^2(N_1-N_0)(1-\psi^3)(1-\psi)[N_0(\psi^2+1)-(3N_1-N_0)\psi] - 27N_2N_1^3\psi^3 = 0 \quad ,$$

a reciprocal sextic polynomial, which reduces in the usual way to

$$az^3 + bz^2 + cz + d = 0 \quad ,$$

where

$$a = N_0(N_2-N_1)^2(N_1-N_0) \quad ,$$
$$b = -(N_2-N_1)^2(N_1-N_0)(3N_1+2N_0) \quad ,$$
$$c = (N_2-N_1)^2(N_1-N_0)(3N_1-N_0) \quad ,$$
$$d = 2(N_2-N_1)^2(N_1-N_0)(3N_1+N_0) - 27N_2N_1^3 \quad . \tag{5.6.3}$$

We denote these critical values by $\tilde{\psi}_0$ and note that they correspond to the case when $t_0 = \tilde{t}$.

The poles of t, (5.6.1), are obtained in exactly the same way. The poles of t_0 and the zeros of q, all calculated in Sect.5.2,3, provide one set of poles; additional poles are obtained by setting the denominator equal to zero:

$$N_2f(N_0-q) + t_0(N_1-q) = 0 \quad . \tag{5.6.4}$$

By treating this in the same way, we get the reciprocal sextic in

$$(N_2-N_1)^2(N_1-N_0)(1-\psi)(1-\psi^3)[N_0(\psi^2+1)+(3N_1+N_0)\psi] + 27N_2N_1^3\psi^3 = 0 \quad ,$$

which reduces, in the usual way, to the cubic

$$az^3 + bz^2 + cz + d = 0 \quad,$$

where

$$a = N_0(N_2-N_1)^2(N_1-N_0) \quad,$$

$$b = 3N_1(N_2-N_1)^2(N_1-N_0) \quad;$$

$$c = -3(N_2-N_1)^2(N_1^2-N_0^2) \quad,$$

$$d = -2(N_2-N_1)^2(N_1-N_0)(3N_1+N_0) + 27N_2N_1^3 \quad. \tag{5.6.5}$$

We denote these critical values by $\tilde{\psi}_\infty$ and $\tilde{\phi}_\infty$.

Finally, we consider the case in which $\tilde{t} = t_1$. By using (5.6.1) [or (4.4.11)] and (3.1.9), we obtain

$$\frac{t_0}{q} \frac{fN_2[N_1q+N_0(N_1-2q)] - N_1t_0(N_1-q)}{N_2f(N_0-q) + t_0(N_1-q)} = \frac{N_1(N_0f-t_0)}{q} \quad. \tag{5.6.6}$$

When running through manipulative exercises, such as those encountered in this chapter, it is essential to be alert to the possible occurrence of what I call *nice things*. In every preceding chapter there has been at least one nice thing, that is, an easy way out of a complicated calculation — some thing that factors in an interesting or useful way, or some fortunate event that breaks up the monotony. Until now, there has been a dearth of nice things in this chapter. But when we go through the calculations involved in simplifying (5.6.6), we find that it becomes

$$t_0(N_1-2N_2)(N_1-q) + N_2N_1f(N_0-q) = 0$$

in a delightful manner, and moreover is identical to (5.5.8), which has already been treated.

This should really have been expected. When the exit pupil and the second surface coincide, then the inner stop and the second surface must also coincide.

5.7 Zero Petzval Contribution

The rest of this chapter is concerned with another type of critical value.
These are associated with the aberrations themselves; in particular, they
are the values of the shape parameters at which an aberration or a group of
aberrations vanishes. The first to be considered is the Petzval sum. This
type of singularity was studied most extensively by MERCADO [5.2]; the account presented here is based on his work.

The procedure followed is identical to that pursued in tracking down the
other critical values. We begin with the expression for the Petzval sum for
a module, (4.6.1):

$$p = [N_2 f(N_0-q)+t_0 q]/N_2 N_1 N_0 f t_0 \quad . \tag{5.7.1}$$

An obvious zero for p occurs at a pole for t_0. Others are obtained by solving
the equation

$$N_2 f(N_0-q) + t_0 q = 0 \quad , \tag{5.7.2}$$

after the formulas for q and t_0, from (4.3.14,15), are substituted in it.
This leads, in turn, to the reciprocal octic polynomial

$$(N_2-N_1)^2(N_1-N_0)^2(1-\psi^3)^2[N_0(\psi^2+1)+(3N_1+N_0)\psi] - 81N_2 N_1^4 \psi^4 = 0 \quad .$$

After the substitution given in (5.5.5), this reduces to the quartic in z,

$$(N_2-N_1)^2(N_1-N_0)^2(z^3-3z-z)(N_0 z+3N_1+N_0) - 81N_2 N_1^4 = 0 \quad ,$$

or, expanded out,

$$az^4 + bz^3 + cz^2 + dz + e = 0 \quad ,$$

where

$$a = N_0(N_2-N_1)^2(N_1-N_0)^2 \quad ,$$

$$b = (N_2-N_1)^2(N_1-N_0)^2(3N_1+N_0) \quad ,$$

$$c = -3N_0(N_2-N_1)^2(N_1-N_0)^2 \quad ,$$

$$d = -(N_2-N_1)^2(N_1-N_0)^2(9N_1+5N_0) \quad ,$$

$$e = -2(N_2-N_1)^2(N_1-N_0)^2(3N_1+N_0) - 81N_2N_1^4 \quad . \tag{5.7.3}$$

MERCADO [5.2] did not stop here in his analysis. By a very clever application of the theorem of Budan and Fourier (see, for example, [5.3]) and use of Sturm's functions (see, for example, [5.4] or [5.5]), he was able to determine the number of zeros of (5.7.3) and their approximate locations. A summary of his results follows.

Consider, first, refracting modules in which the refractive indices are either unity or have positive values in, roughly the range 1.4-2.0. For the imaginary case, Mercado showed that the Petzval sum for a module never vanishes. For the real case, he found that the Petzval sum vanishes for two values of ψ. One of these zeros lies in the interval $(-0.267949,0)$. The location of the second root varies with the sign of (N_1-N_0). When the sign of this quantity is positive, the zero lies in the interval $(0,0.208712)$; when it is negative the zero is in $(0,0.381966)$.

Mercado also studied reflecting systems in which $N_2 = N_0$ and $N_1 = -N_0$. He found that two zeros also occurred; one in the real case and one for the imaginary case. The real-case zero occurred at $\psi = 0.46341164$; the imaginary-case zero at $\phi = 0.19917$.

We will refer to this class of singularities as ψ_p and ϕ_p.

5.8 The Concentric Situation

The concentric lens is a very old chestnut. It has the very desirable property that all asymmetric aberrations are zero. It also has the very undesirable property that field curvature dominates all other properties of the lens. The concentric lens, nevertheless, is important to *modulophiles*. It turns out that modules become concentric for at least one value of the shape parameter. ANDERSON [5.6], who has studied the concentric situation most thoroughly, argues that this value of the shape parameter is one of the two better places to initiate the design process, and that most useful designs are found in its neighborhood. However, this is not the place to begin this discussion.

We begin with (4.4.6), the condition for concentricity:

$$N_2f(N_0-q) - t_0(N_1-q) = 0 \quad . \tag{5.8.1}$$

We follow slavishly the procedures established previously, substituting for q and t_0 from (4.3.14,15), and obtain the reciprocal octic polynomial

$$(N_2-N_1)^2(N_1-N_0)^2(1-\psi^3)^2[N_0(\psi^2+1)+(3N_1+N_0)\psi]$$
$$- 27N_2N_1\psi^3[(N_1-N_0)(\psi^2+1)+(4N_1-N_0)\psi] = 0 \quad . \tag{5.8.2}$$

Once more, we replace ψ by z, making use of (5.5.5), and obtain

$$(N_2-N_1)^2(N_1-N_0)^2(z^3-3z-2)(N_0z+3N_1+N_0) - 27N_2N_1^3[(N_1-N_0)z+4N_1-N_0] = 0 \quad ,$$

which, when expanded, becomes

$$az^4 + bz^3 + cz^2 + dz + e = 0 \quad ,$$

where

$$a = N_0(N_2-N_1)^2(N_1-N_0)^2 \quad ,$$
$$b = (N_2-N_1)^2(N_1-N_0)^2(3N_1+N_0) \quad ,$$
$$c = -3N_0(N_2-N_1)^2(N_1-N_0)^2 \quad ,$$
$$d = -(N_2-N_1)^2(N_1-N_0)^2(9N_1+5N_0) - 27N_2N_1^3(N_1-N_0) \quad ,$$
$$e = -2(N_2-N_1)^2(N_1-N_0)^2(3N_1+N_0) - 27N_2N_1^3(4N_1-N_0) \quad . \tag{5.8.3}$$

We will designate the members of this class of singularities by the symbols ψ_c and ϕ_c.

5.9 The Quartic Polynomial

We have seen several cases in which we should have or could have used algebraic solutions of cubic and quartic polynomials, but we have not done so mainly because the process was too difficult. That is not to say that the task is impossible. Indeed it is very possible that formal solutions for critical values are not only obtainable, albeit at the expense of considerable effort, but also, once obtained, they might be of great value. For this reason, the following discussion is included.

The quartic and its solution are closely associated with Cardano, whom we have already encountered in our discussion of the cubic, and his companion Lodovico Ferrari (1522-1565). Cardano hired Ferrari as a servant when the latter was about 14 years old. When Cardano found that Ferrari could read and write, he was promoted to the position of secretary, in which job he assisted with the preparation of the manuscript of *Arithmatica, Geometrica, Musica*. During this time, Ferrari learned enough about mathematics to do original work. Indeed, he was about twenty years old when he obtained the solution of the quartic [5.7].

A numerical problem whose solution depended on solving a quartic was presented to Cardano by Zuanne de Tonini da Coi, who asserted that it was insolvable. Cardano disagreed and turned the problem over to Ferrari. The solution of the cubic, on which the quartic solution depends, was known to both Ferrari and Cardano, but not necessarily to da Coi. Ferrari succeeded and his general solution appeared in Cardano's next book on algebra, *Ars Magna*, in which he gave Ferrari full credit. The method has hardly been improved upon in the intervening 400 years [5.8].

ORE [5.7, pp.80-81] described Ferrari in these words, "The renaissance abounds in impulsive and hot-headed geniuses and Ferrari ran true to form. He had such a temper that even Cardano at times was afraid to speak to him, and one day when he was seventeen years old he came home from a brawl missing the fingers of his right hand." It has been said that at the age of 43 he died by poison, probably by the hand of either his sister or her lover.

Consider the general quartic polynomial

$$ax^4 + bx^3 + cx^2 + dx + e = 0 \quad . \tag{5.9.1}$$

Because the leading coefficient cannot be zero, we may, with no loss of generality, divide through by it. Also, as in the case of the cubic polynomial, we may transform the variable in such a way as to cause the cubic term to drop out, thus obtaining the reduced quartic

$$x^4 + cx^2 + dx + e = 0 \quad . \tag{5.9.2}$$

We introduce a variable parameter p and rewrite this equation in the form

$$x^4 + (cp)x^2 + \frac{1}{4}(c+p)^2 = px^2 - dx + \frac{1}{4}(c+p)^2 - e \quad .$$

Note that the left-hand member of this expression is a perfect square, so we may write

$$\left[x^2 + \frac{1}{2}(c+p)\right]^2 = px^2 - dx + \frac{1}{4}(c+p)^2 - e \quad . \tag{5.9.3}$$

The strategy that we will use here is to find a value of p for which the right-hand member of (5.9.3) is also a perfect square. Then, by taking the square root of both sides, we obtain a quadratic polynomial, which is easily solved.

Now the right-hand member of (5.9.3) is also a quadratic in x. For it to be a perfect square, it is necessary and sufficient that its discriminant vanish. Setting the discriminant equal to zero results in a cubic polynomial in the variable p,

$$p^3 + 2cp^2 + (c^2-4e)p - d^2 = 0 \quad , \tag{5.9.4}$$

referred to as the *resolvent cubic*. This we know how to solve.

Now let p_1, p_2, and p_3 be the three roots of (5.9.4), the resolvent cubic. It is well known (see, for example, [5.9]) that the symmetric functions of the roots of a polynomial are proportional to its coefficients, so that

$$
\begin{aligned}
p_1 + p_2 + p_3 &= -2c \quad , \\
p_1 p_2 + p_1 p_3 + p_2 p_3 &= c^2 - 4e \quad , \\
p_1 p_2 p_3 &= c_2 \quad .
\end{aligned}
\tag{5.9.5}
$$

Next, substitute one of these roots, say p_1, into the quartic in the form given in (5.9.3) and replace the coefficients by the expressions in (5.9.5). The result is

$$\left[x^2 + \frac{1}{4}(p_1 - p_2 - p_3)\right]^2 = \left(\sqrt{p_1}\,x - \frac{1}{2}\sqrt{p_2 p_3}\right)^2 \quad .$$

Note that the right-hand member is a perfect square, exactly as it should be.

By taking the square root of both sides of this equation, we obtain two quadratic polynomials in x, corresponding to the two branches of the square root. They are

$$x^2 + \frac{1}{4}(p_1 - p_2 - p_3) = \sqrt{p_1}\,x - \frac{1}{2}\sqrt{p_2 p_3} \quad ,$$

$$x^2 + \frac{1}{4}(p_1-p_2-p_3) = -\sqrt{p_1}\,x + \frac{1}{2}\sqrt{p_2p_3} \quad . \tag{5.9.6}$$

Rearranging the terms of the first of these, we can easily obtain

$$x^2 + \sqrt{p_1}\,x + \frac{1}{4}\left[p_1 - (\sqrt{p_2}-\sqrt{p_3})^2\right] = 0 \quad ,$$

whose two solutions are

$$x = \frac{1}{2}\left[\sqrt{p_1} \pm (\sqrt{p_2}-\sqrt{p_3})\right] \quad .$$

We do the same thing with the second quadratic in (5.9.6) and get

$$x = \frac{1}{2}\left[-\sqrt{p_1} \pm (\sqrt{p_2}+\sqrt{p_3})\right] \quad .$$

This result is rather amazing. It states that the four roots of the quartic are equal to certain sums and differences of the roots of its resolvent cubic,

$$\begin{aligned}
x_1 &= (\sqrt{p_1}+\sqrt{p_2}-\sqrt{p_3})/2 \quad , \\
x_2 &= (\sqrt{p_1}-\sqrt{p_2}+\sqrt{p_3})/2 \quad , \\
x_3 &= (-\sqrt{p_1}+\sqrt{p_2}+\sqrt{p_3})/2 \quad , \\
x_4 &= (-\sqrt{p_1}-\sqrt{p_2}-\sqrt{p_3})/2 \quad .
\end{aligned} \tag{5.9.7}$$

6. The Canonical Equations

In the previous chapter, we examined the functions q and t_0 fairly carefully. We looked at the critical values of their arguments and put together a preliminary description of a classification scheme for constructing some necessary conditions for successfully coupling modules. In this chapter, we complete the classification scheme. However, the most important step is the definition of canonical optical parameters, canonical ray tracing, and canonical aberration coefficients. This enables us not only to couple modules but to assemble strings of coupled modules in which certain third-order aberrations and primary chromatic aberrations are set, a priori, equal to zero. The work described in this chapter is based on the results of MERCADO [6.1,2] and ANDERSON [6.3].

6.1 The Canonical Parameters

We begin by defining the canonical module function for the real case from (4.3.14),

$$Q = \frac{-3\psi}{(1-\mu_1)(\psi^2+1) - (2+\mu_1)\psi} \quad ,$$
(6.1.1)

and for the imaginary case from (4.3.17),

$$Q = \frac{3}{(2+\mu_1) + 2(1-\mu_1)\cos\phi} \quad .$$
(6.1.2)

We adopt the convention that all canonical quantities will be designated by capital letters. Note that from (4.3.14,17) we can obtain

$$q = N_1 Q \quad .$$
(6.1.3)

We next define a canonical counterpart for t_0. For the real case, from (4.3.15), we denote

$$T_0 = \frac{27N_1\psi^3}{(1-\mu_2)^2(1-\mu_1)(2-\psi^3)^2} \quad , \tag{6.1.4}$$

and for the imaginary case, using (4.3.18),

$$T_0 = \frac{-27N_1}{2(1-\mu_2)^2(1-\mu_1)(1+\cos3\phi)} \quad . \tag{6.1.5}$$

The cubic polynomial (4.1.4) now becomes

$$T_0(1-\mu_2)^2(1-Q)^2(1-\mu_1Q) + (1-\mu_1)^2Q^3 = 0 \quad . \tag{6.1.6}$$

Next, using (3.1.8-10), we define

$$C_1 = \frac{\mu_1-Q}{(1-\mu_1)T_0} \quad , \tag{6.1.7}$$

$$T_1 = \frac{N_0-T_0}{Q} \quad , \tag{6.1.8}$$

$$C_2 = \frac{Q}{N_2\mu_1(1-\mu_2)} \quad , \tag{6.1.9}$$

which we follow with definitions of

$$E = \frac{2N_2T_0Q(1-Q)}{N_2(\mu_1-Q)(2Q-1) + T_0(1-Q)} \quad , \tag{6.1.10}$$

$$A = -N_0\frac{T_0(\mu_2-2)(1-Q) + N_1(\mu_1-Q)}{2T_0\mu_2Q(1-Q)} \quad , \tag{6.1.11}$$

$$G = \frac{T_0[N_2Q+\mu_1(1-2Q)-T_0(1-Q)]}{Q[N_2(\mu_1-Q)+T_0(1-Q)]} \quad . \tag{6.1.12}$$

The latter follow from (4.4.9-11).

It follows that

$$t_0 = fT_0 \quad , \quad t_1 = fT_1 \quad , \quad \bar{t} = fE \quad , \quad \bar{t}' = fA \quad , \quad \tilde{t} = fG \quad ,$$
$$c_1 = C_1/f \quad , \quad c_2 = C_2/f \quad , \quad q = N_1Q \quad . \tag{6.1.13}$$

To minimize confusion, we will refer to the parameters t_0, t_1, c_1, etc. as the *optical parameters*, and we will call T_0, T_1, C_1, etc. the *canonical parameters*. The important thing to note is that the canonical parameters do not depend on the power parameter, but only on the shape parameter.

6.2 Cardinal Points

It is useful to look at the canonical versions of the cardinal points developed in Sect.3.2. We make use of (3.2.4,7,8) as well as (6.1.3,13). We obtain for the first and second foci, respectively,

$$T_0 \quad , \quad N_2 M_1 T_0 - N_1(\mu_1 - Q)T_0 Q \quad . \tag{6.2.1}$$

As always, the second focus is measured from the second surface. The canonical principal planes are obtained from (3.2.7) and are

$$D_0 = N_0 \quad , \quad D_2 = N_2(\mu_1 - Q)(N_0 - Q)/T_0 Q \quad , \tag{6.2.2}$$

whereas the canonical nodal planes come from (3.2.8) and are

$$D_0 = N_2 \quad , \quad D_2 = \frac{N_2 \mu_1}{T_0 Q} T_0 - N_1(\mu_1 - Q) - \mu_2 T_0 Q \quad . \tag{6.2.3}$$

Note that the first principal plane and the first nodal plane are measured from the main focus, whereas those on the second side are measured from the second surface.

6.3 Canonical Ray Tracing

Canonical ray tracing introduces some further abstractions. We define the canonical ray variables recursively, following the paraxial ray-tracing equations given in (1.2.1,2,5):

$$Y_{j+1} = Y_j + T_j U_j \quad ,$$
$$U_{j+1} = \mu_{j+1} U_j - (1-\mu_{j+1})C_{j+1} Y_{j+1} \quad . \tag{6.3.1}$$

We distinguish marginal canonical rays from chief canonical rays. We also require the counterpart of the refraction invariant, defined in (1.2.4),

$$A_{j+1} = N_{j+1}(C_{j+1}Y_{j+1}+U_{j+1}) = N_j(C_{j+1}Y_{j+1}+U_j) \quad . \tag{6.3.2}$$

For the canonical marginal ray, we set $Y_0 = 0$ and $U_0 = 1/N_0$. Then, from (6.3.1,2) as well as (6.1.7-9) and (6.1.10-12) we obtain

$$
\begin{aligned}
&Y_0 = 0 \;, \qquad U_0 = 1/N_0 \;, \\
&Y_1 = T_0/N_0 \;, \quad U_1 = Q/N_0 \;, \quad A_1 = (1-Q)/(1-\mu_1) \\
&Y_2 = 1 \;, \qquad U_2 = 0 \;, \qquad A_2 = Q/\mu_1(1-\mu_2) \quad .
\end{aligned}
\tag{6.3.3}
$$

We do very much the same thing for the canonical chief ray. We set $\bar{Y}_0 = 1$. Because this ray must pass through the axial point of the pupil plane, its slope must be $\bar{U}_0 = -1/E$. Again using (6.3.1,2), (6.1.7,9), and (6.1.10,12), we obtain

$$
\begin{aligned}
\bar{Y}_0 &= 1 \;, & \bar{U}_0 &= -1/E \;, \\
\bar{Y}_1 &= 1 - T_0/E \;, & \bar{U}_1 &= -(\mu_1 E+T_0 Q-EQ)/T_0 E \;, \\
& & \bar{A}_1 &= \frac{N_0}{(1-\mu_1)T_0 E} [-T_0(1-Q+E(\mu_1-Q)] \;, \\
\bar{Y}_2 &= \frac{1}{T_0 EQ} & \bar{U}_2 &= -\frac{1}{N_2} \;, \\
&\times[N_0 Q(E-T_0)-\mu_1 E(N_0-T_0)] \;, & & \\
& & \bar{A}_2 &= \frac{1}{\mu_1(1-\mu_2)T_0 E} \\
& & &\times[N_0 EQ-N_0 T_0 Q-N_0\mu_1 E+\mu_1\mu_2 T_0 E] \quad .
\end{aligned}
\tag{6.3.4}
$$

The canonical counterpart of the Lagrange invariant is

$$H = N(\bar{U}Y-U\bar{Y}) \equiv 1 \;, \tag{6.3.5}$$

as can be seen from the preceeding equations.

It is convenient to arrange the preceeding equations so that the chief-ray quantities are expressed in terms of the marginal-ray quantities. Thus

$$\bar{Y}_0 = -\frac{N_0}{E} Y_0 + 1 \;,$$

$$\bar{U}_0 = -\frac{N_0}{E} U_0 \quad ,$$

$$\bar{Y}_1 = -\frac{N_0}{E} Y_1 + 1 \quad ,$$

$$\bar{U}_1 = -\frac{N_0}{E} U_1 - \frac{\mu_1 - Q}{T_0} \quad ,$$

$$\bar{A}_1 = -\frac{N_0}{E} A_1 + \frac{N_0(\mu_1 - Q)}{(1-\mu_1)T_0} \quad ,$$

$$\bar{U}_2 = -\frac{N_0}{E} U_2 - \frac{1}{N_2} \quad ,$$

$$\bar{Y}_2 = -\frac{N_0}{E} Y_2 - \frac{N_0(\mu_1 - Q) - \mu_1 T_0}{T_0 Q} \quad ,$$

$$\bar{A}_2 = -\frac{N_0}{E} A_2 - \frac{N_0(\mu_1 - Q) - \mu_1 \mu_2 T_0}{\mu_1(1-\mu_2)T_0} \quad . \tag{6.3.6}$$

6.4 Canonical Aberration Coefficients

Now we come to the derivation of the canonical version of the aberration contributions obtained in earlier chapters. We have by this time run out of symbols, so where there is danger of ambiguity we will use the carat symbol to denote the canonical variable. We begin with the expression for α, β, γ, and δ found in (3.3.4,4). First, we treat α_1 and apply (6.1.13):

$$\alpha_1 = \mu_1(N_1 - q)/(1-\mu_1) = N_0(1-Q)/(1-\mu_1) \quad . \tag{6.4.1}$$

In like manner, we find that

$$\alpha_2 = N_1 Q/(1-\mu_2) \quad ,$$
$$\gamma_1 = (1-\mu_1 Q)/N_0 \quad ,$$
$$\gamma_2 = Q/N_1 \quad . \tag{6.4.2}$$

The remaining quantities in this collection require a little more care,

$$\beta_1 = \mu_1(N_0 - q)/(1-\mu_1)t_0 = N_0(\mu_1 - Q)/f(1-\mu_1)T_0 \quad .$$

By a series of similar calculations, we find the remaining quantities,

$$\beta_1 = \hat{\beta}_1/f \quad , \quad \hat{\beta}_1 = N_0(\mu_1-Q)/(1-\mu_1)T_0 \quad ,$$

$$\beta_2 = \hat{\beta}_2/f \quad , \quad \hat{\beta}_2 = [\mu_2 T_0 - N_1(\mu_1-Q)]/(1-\mu_2)T_0 \quad ,$$

$$\delta_1 = \hat{\delta}_1/f \quad , \quad \hat{\delta}_1 = (\mu_1-Q)/N_1 T_0 \quad ,$$

$$\delta_2 = \hat{\delta}_2/f \quad , \quad \hat{\delta}_2 = [\mu_2 T_0 - N_2(\mu_1-Q)]/N_1 T_0 \quad . \tag{6.4.3}$$

We do the same sort of thing with the expressions defined in (3.3.11) and obtain

$$U_1 = f\hat{U}_1 \quad , \quad \hat{U}_1 = T_0\gamma_1\alpha_1^2 + N_0\gamma_2\alpha_2^2 \quad ,$$

$$U_2 = \hat{U}_2 \quad , \quad \hat{U}_2 = T_0\gamma_1\alpha_1\hat{\beta}_1 + N_0\gamma_2\alpha_2\hat{\beta}_2 \quad ,$$

$$U_3 = \hat{U}_3/f \quad , \quad \hat{U}_3 = T_0\gamma_1\hat{\beta}_1^2 + N_0\gamma_2\hat{\beta}_2^2 \quad . \tag{6.4.4}$$

Of course, this is only a formal development — $\hat{U}_1$ remains equal to zero — and from (4.4.7,8) and the expression for $\bar{t}$ in (6.1.13) we obtain

$$2\hat{U}_2 - E\hat{U}_3 = 0 \quad . \tag{6.4.5}$$

Next, we scrutinize the quantity A defined in (3.2.2):

$$A = N_0[t_0 - f(N_0-q)]/t_0 q = \mu_1[T_0 - N_1(\mu_1-Q)]/T_0 Q \quad . \tag{6.4.6}$$

This series of preliminary calculations concludes with the canonization (without necessarily the beatification) of the quantities defined in (3.4.2,4):

$$W_1 = \hat{W}_1 \quad , \quad \hat{W}_1 = T_0\alpha_1^2\hat{\delta}_1 + N_0\alpha_2^2\hat{\delta}_2 \quad ,$$

$$W_2 = \hat{W}_2 \quad , \quad \hat{W}_2 = \alpha_1^2\gamma_1 + \alpha_2^2 A\gamma_2 \quad ,$$

$$W_3 = \hat{W}_3/f \quad , \quad \hat{W}_3 = \alpha_1^2\hat{\delta}_1 + \alpha_2^2 A\hat{\delta}_2 \quad ; \tag{6.4.7}$$

$$V_1 = \hat{V}_1/f \quad , \quad \hat{V}_1 = T_0\alpha_1\hat{\beta}_1\hat{\delta}_1 + N_0\alpha_2\hat{\beta}_2\hat{\delta}_2 \quad ,$$

$$V_2 = \hat{V}_2/f \quad , \quad \hat{V}_2 = \alpha_1\hat{\beta}_1\gamma_1 + \alpha_2\hat{\beta}_2 A\gamma_2 \quad ,$$

$$V_3 = \hat{V}_3/f^2 \quad , \quad \hat{V}_3 = \alpha_1\hat{\beta}_1\hat{\delta}_1 + \alpha_2\hat{\beta}_2 A\hat{\delta}_2 \quad ; \tag{6.4.8}$$

$$V_4 = \hat{V}_4/f^2 \quad , \quad \hat{V}_4 = T_0\hat{\beta}_1^2\hat{\delta}_1 + N_0\hat{\beta}_2^2\hat{\delta}_2 \quad ,$$

$$V_5 = \hat{V}_5/f^2 \quad , \quad \hat{V}_5 = \hat{\beta}_1^2\gamma_1 + \hat{\beta}_2^2 A\gamma_2 \quad ,$$

$$V_6 = \hat{V}_6/f^2 \quad , \quad \hat{V}_6 = \hat{\beta}_1^2\hat{\delta}_1 + \hat{\beta}_2^2 A\hat{\delta}_2 \quad . \tag{6.4.9}$$

Now we come to the calculation of the aberration coefficients. First, we need to refer to (6.3.3,4) to define a relationship between the real-ray variables and their canonical counterparts. We adopt the formal convention,

$$\bar{y}_0 = \bar{y}_0 \Upsilon_0 \quad ,$$
$$u_0 = u_0 N_0 U_0 \quad , \quad \bar{u}_0 = \bar{U}_0 \bar{y}_0 / f \quad , \tag{6.4.10}$$

so that the Lagrange invariant, first defined in (1.2.9), becomes

$$h = N_0 \bar{y}_0 u_0 H \quad , \tag{6.4.11}$$

where H is the canonical version defined in (6.3.5) and is identically equal to unity.

We also need expressions for the ray variables on the output side of the module. These are

$$y_2 = N_0 u_0 f \Upsilon_2 \quad , \quad \bar{y}_2 = \bar{y}_0 \Upsilon_2 \quad ,$$
$$\bar{u}_2 = (\bar{y}_0 / f) \bar{U}_2 \quad . \tag{6.4.12}$$

The nonzero image error is coma, which we find in (4.4.2):

$$\beta = u_0^3 \bar{y}_0 U_2 \quad .$$

Substituting from (6.4.4,11) gives us

$$\beta = u_0^3 N_0^3 \bar{y}_0 F \quad , \quad F = U_0^3 \Upsilon_0 \hat{U}_2 \quad . \tag{6.4.13}$$

The next items in the program are the calculation of canonical pupil astigmatism, from (4.5.1),

$$\bar{c} = u_0^2 N_0^2 \bar{y}_0^2 \bar{c} / f \quad , \quad \bar{c} = U_0^2 \Upsilon_0 [\bar{U}_0 (\hat{W}_1 + \hat{W}_2) + \Upsilon_0 \hat{W}_3] \quad ,$$
$$\bar{c} = u_0^2 N_0^2 \bar{y}_0^2 \bar{c} / f \quad ; \tag{6.4.14}$$

canonical pupil coma, from (4.5.3),

$$\beta = u_0 N_0 \bar{y}_0^3 \bar{F} / f^2 \quad ,$$
$$\bar{F} = F(\bar{U}_0 / U_0)^2 + C(\bar{U}_0 / U_0) + U_0 \bar{y}_0^2 [\bar{U}_0 (\hat{V}_1 + \hat{V}_2) + \Upsilon_0 \hat{V}_3] \quad ; \tag{6.4.15}$$

and canonical pupil spherical aberration, from (4.5.5),

$$\beta = \bar{y}_0^{-4}\bar{B}/f^3 \quad,$$

$$\bar{B} = -2F(\bar{U}_0/U_0)^3 - \bar{C}(\bar{U}_0/U_0)^2 + 2\bar{F}(\bar{U}_0/U_0) + \bar{y}_0^3[U_0(\hat{V}_4+\hat{V}_5)+\bar{y}_0\hat{V}_6] \quad. \quad (6.4.16)$$

The calculation of the field errors come next in this section. The Petzval contribution comes easiest from (3.5.1), into which we substitute the appropriate quantities from (6.1.13) and obtain

$$p = P/f \quad, \quad P = [N_2(\mu_1-Q)+T_0Q]/N_2N_0T_0 \quad. \quad (6.4.17)$$

Canonical image distortion is gotten from (3.5.2)

$$e = \bar{y}_0^3 u_0 N_0 E/f^2 \quad, \quad E = \bar{F} + H(\bar{U}_0^2-\bar{U}_0^2) \quad, \quad (6.4.18)$$

and canonical pupil distortion is gleaned from (3.5.3),

$$\bar{e} = -u_0^3 N_0^3 \bar{y}_0 \bar{E} \quad, \quad \bar{E} = F + H(U_0^2-u_0^2) \quad. \quad (6.4.19)$$

The final enterprise in this section concerns the chromatic aberrations, which we last encountered in (4.6.4-7). There, we saw that longitudinal chromatic aberration could be written in the form $\ell = u_0^2\hat{L}$, where

$$\hat{L} = \alpha_1 t_0(D_1/N_1+D_0/N_0) + N_0 f\alpha_2(D_2/N_2-D_1/N_1) \quad.$$

Making the appropriate substitutions from (3.3.4), we find

$$\ell = u_0^2 N_0^2 f \mathscr{L} \quad, \quad \mathscr{L} = U_0^2\hat{L} \quad, \quad (6.4.20)$$

where

$$\hat{L} = T_0(1-Q)(\mu_1 D_1-D_0)/(1-\mu_1) + N_0 Q(\mu_2 D_2-D_1)/(1-\mu_2)$$
$$= \alpha_1 T_0(D_1/N_1-D_0/N_0) + N_0\alpha_2(D_2/N_2-D_1/N_1) \quad. \quad (6.4.21)$$

Transverse chromatic aberration is given by $\ell = u_0(\bar{u}_0\hat{L}+\bar{y}_0\hat{T})$, where

$$\hat{T} = t_0\beta_1(D_1/N_1-D_0/N_0) + N_0 f\beta_2(D_2/N_2-D_1/N_1) \quad.$$

We find that, from (3.3.4) and (6.4.10), this becomes

$$t = u_0 N_0 \bar{y}_0 \mathcal{T} \quad , \quad \mathcal{T} = U_0 (\bar{U}_0 \hat{L} + \bar{V}_0 \hat{T}) \quad , \tag{6.4.22}$$

where

$$\hat{T} = (\mu_1 - Q)(\mu_1 D_1 - D_0)/(1-\mu_1) + \mu_1 \mu_2 [T_0 - N_2 (\mu_1 - Q)](\mu_2 D_2 - D_1)/(1-\mu_2 T_0)$$
$$= T_0 \hat{\beta}_1 (D_1/N_1 - D_0/N_0) + N_0 \hat{\beta}_2 (D_2/N_2 - D_1/N_1) \quad . \tag{6.4.23}$$

We also need to define the canonical versions of the intermediate third-order quantities s_j and $\bar{s}_j$, first given in (1.2.11) and encountered in terms of module quantities in (3.3.5). They will be required in the calculation of the fifth-order aberration contributions. We make use of the definition of certain canonical quantities in (6.4.1-3). From these two sets of equations, we may write

$$s_j = u_0 \gamma_j \quad , \quad \bar{s}_j = -\bar{u}_0 \gamma_j - \bar{y}_0 \delta_j / f \quad , \quad j = 1,2 \quad .$$

By substitution from (6.4.10), these become

$$s_j = u_0 N_0 S_j \quad , \quad S_j = -U_0 \gamma_j \quad ,$$
$$\bar{s}_j = (\bar{y}_0/f)\bar{S}_j \quad , \quad \bar{S}_j = -\bar{U}_0 \gamma_j - \bar{V}_0 \hat{\delta}_j \quad , \quad j = 1,2 \quad . \tag{6.4.24}$$

These equations are in many respects disappointing and unsatisfactory. In each, it is possible to do a considerable amount of cancellation and reduction, by applying repeatedly (6.4.5) and the fact that $U_2 = 0$. The result of all this is a disheartening hodge-podge of formulas, not appreciably simpler than the ones given, and without any evident underlying structure to make programming simpler and comprehension easier.

6.5 The Forward Orientation

We are now in a position to define canonical versions of the aberration coefficients. At the same time we can, for the first time, visualize the module as a component of an optical system. This is illustrated in Fig.6.1. This shows a forward module as a constituent part of a lens, where its first sur-

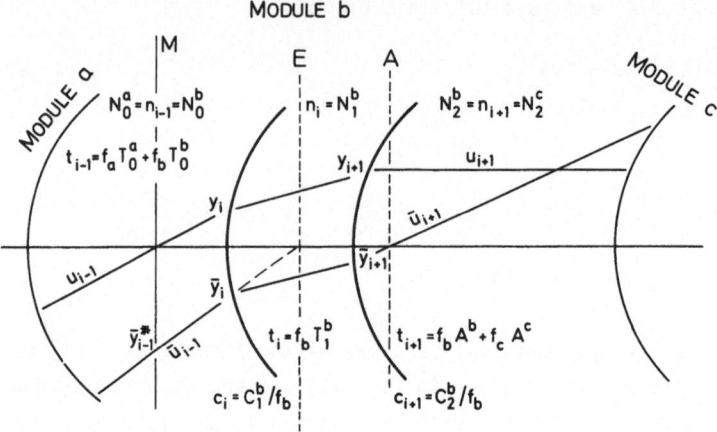

Fig. 6.1. The module as a component — the forward orientation

Table 6.1. Forward orientation — optical parameters

$n_{i-1} = N_0$	$c_i = C_1/f$	$t_{i-1} = \ldots + fT_0$
$n_i = N_1$	$t_i = fT_1$	$= \ldots + fE$
$n_{i+1} = N_2$	$c_{i+1} = C_2/f$	$t_{i+1} = fA + \ldots$
$\bar{t} = fE$	$\bar{t}' = fA$	
$p = P/f$		

face is the system's i^{th} surface. Table 6.1 shows the relations between the canonical optical parameters, the lens parameters, and f, the power parameter. Also included in this table is a calculation of the Petzval sum. We make use of (3.5.1) and (6.1.14-16).

We use these identities to establish relationships between the ray-traced variables and the results of canonical ray tracing. It is perhaps simplest to regard the module as a device that receives paraxial rays from the left-hand side as input, and emits the same sort of rays on the right-hand side as output. On the plane of main focus, we discover a marginal ray defined by a slope u_{i-1} and which passes through the axis. We may write formally $u_{i-1} = u_{i-1} n_{i-1} U_0$ because $n_{i-1} U_0 = N_0 U_0 = 1$. Applying the ray-tracing equations, (1.2.1-5) and (6.3.3), we obtain a list of variables for the marginal ray traced through the modules. This list is found in Table 6.2.

The same thing is done for the chief ray. Here we need to specify $\bar{y}_{i-1}^{\#}$, the height of the chief ray on the plane of the main focus. Then, again making

Table 6.2. Forward orientation — ray relationships

Marginal	Chief
$u_{i-1} = u_{i-1} m_{i-1} U_0$	$\bar{u}_{i-1} = (\bar{y}^{\#}_{i-1}/f)\bar{U}_0$
$y_i = f u_{i-1} n_{i-1} Y_1$	$\bar{y}_i = \bar{y}^{\#}_{i-1} \bar{Y}_1$
$a_i = u_{i-1} n_{i-1} A_1$	$\bar{a}_i = (\bar{y}^{\#}_{i-1}/f)\bar{A}_1$
$u_i = u_{i-1} n_{i-1} U_1$	$\bar{u}_i = (\bar{y}^{\#}_{i-1}/f)\bar{U}_1$
$y_{i+1} = f u_{i-1} n_{i-1} Y_2$	$\bar{y}_{i+1} = \bar{y}^{\#}_{i-1} \bar{Y}_2$
$a_{i+1} = u_{i-1} n_{i-1} A_2$	$\bar{a}_{i+1} = (\bar{y}^{\#}_{i-1}/f)\bar{A}_2$
$u_{i+1} = u_{i-1} n_{i-1} U_2$	$\bar{u}_{i+1} = (\bar{y}^{\#}_{i-1}/f)\bar{U}_2$
$h_i = n_i(\bar{u}_i y_i - u_i \bar{y}_i) = -u_{i-1} n_{i-1} \bar{y}^{\#}_{i-1} H$	

use of the ray-trace equations (1.2.1-5) and (6.3.4), we obtain a similar
list for the chief-ray variables. These are also in Table 6.2. Also included
in this table is a calculation of the Lagrange invariant, based on (1.2.9)
and (6.3.5).

In all of the ray-trace quantities appear the two input data for the marginal and chief rays, u_{i-1} and $\bar{y}^{\#}_{i-1}$, respectively. The power parameter f of
the module also appears in some of the quantities. The remaining quantities,
the canonical variables, are either constant or depend only on the shape parameter of the module.

Now we proceed to the calculation of the third-order aberrations that
arise from the module whose first surface is the ith surface of the optical
system. The quantities that are input to the module are 1) the slope of the
marginal ray, u_{i-1}, and 2) the height of the chief ray on the main focus,
$y^{\#}_{i-1}$. To be complete, let us begin with spherical aberration and refer to
(3.3.8),

$$b_i = u^4_{i-1} U^{i-1}_1 \ ,$$

into which we substitute the appropriate values from Table 6.2 and (6.4.4) to
get

$$b_i = (-u_{i-1} n_{i-1} U_0)^4 f \hat{U}_1 = u^4_{i-1} n^4_{i-1} f U^4_0 \hat{U}_1 \ . \tag{6.5.1}$$

Here we might call $U^4_0 \hat{U}_1$ canonical spherical aberration but, because it is
identically zero, perhaps we ought not to dignify it with a name.

We can repeat this exercise with the expression for coma, in (3.3.9), and for astigmatism in (3.3.10) and obtain, in exactly the same way,

$$b_i = -u_{i-1}^3 n_{i-1}^3 \bar{y}_{i-1}^{\#} U_0^3 (\bar{U}_0 \hat{U}_1 + \bar{Y}_0 \hat{U}_2) \tag{6.5.2}$$

and

$$c_i = \frac{u_{i-1}^2 n_{i-1}^2 \bar{y}_{i-1}^{\#}}{f} U_0^2 (\bar{U}_0^2 \hat{U}_1 + 2\bar{U}_0 \bar{Y}_0 \hat{U}_2 + \bar{Y}_0^2 \hat{U}_3) \quad . \tag{6.5.3}$$

Because $\hat{U}_1 = 0$, the first of these simplifies and, by means of (6.4.13), may be written as

$$b_i = -u_{i-1}^3 n_{i-1}^3 \bar{y}_{i-1}^{\#} U_0^3 \bar{Y}_0 \hat{U}_2 \quad . \tag{6.5.4}$$

In the second of these, we use (6.4.5), as well as the fact that $U_1 = 0$, to show that c_i is identically zero, as it should be.

The remaining calculations proceed from the equations derived in the last section. The first of these is pupil astigmatism, given in (6.4.14). If we identify u_0 in that equation with u_{i-1}, N_0 with n_{i-1}, and $\bar{y}_0$ with $\bar{y}_{i-1}^{\#}$, we see that

$$c_i = u_{i-1}^2 n_{i-1}^2 \bar{y}_{i-1}^{\#2} \bar{C}/f \quad . \tag{6.5.5}$$

The remaining quantities are obtained in the same way, which is left as an exercise for the reader. These results are collected in Table 6.3.

Table 6.3. Forward orientation — aberration contributions

Image coma	$b_i = u_{i-1}^3 n_{i-1}^3 \bar{y}_{i-1}^{\#} F$
Pupil astigmatism	$\bar{c}_i = u_{i-1}^2 n_{i-1}^2 \bar{y}_{i-1}^{\#2} \bar{C}/f$
Pupil coma	$\bar{b}_i = u_{i-1} n_{i-1} \bar{y}_{i-1}^{\#3} \bar{F}/f^2$
Pupil spherical	$\bar{b}_i = \bar{y}_{i-1}^{\#4} \bar{B}/f^3$
Petzval	$p_i = p/f$
Image distortion	$e_i = u_{i-1} n_{i-1} \bar{y}_{i-1}^{\#3} E/f^2$
Pupil distortion	$\bar{e}_i = u_{i-1}^3 n_{i-1}^3 \bar{y}_{i-1}^{\#} \bar{E}$
Longitudinal color	$\ell_i = u_{i-1}^2 n_{i-1}^2 f \mathscr{L}$
Transverse color	$t_i = u_{i-1} n_{i-1} \bar{y}^{\#} \mathscr{T}$

One last detail remains. We mentioned earlier that u_{i-1} and $\bar{y}_{i-1}$ were to be the input ray-tracing values to the forward module. What, then, shall the output values be? Again, refer to Table 6.2 and (6.4.12) and see that $y_{i+1} = fu_{i-1}n_{i-1}Y_2 = -fu_{i-1}n_{i-1}$ for the marginal ray and for the chief ray $\bar{u}_{i+1} = (\bar{y}_{i-1}^{\#}/f)\bar{U}_2 = -\bar{y}_{i-1}^{\#}/fN_2 = -\bar{y}_{i-1}^{\#}/n_{i+1}$. If this module is followed by a backward module, then these quantities are its inputs. This follows, in the case of the marginal ray, from the fact that $u_{i+1} = 0$, so that $y_{i+1} = y_{i+2}$. These results are collected in Table 6.4.

Table 6.4. Forward orientation — module input and output

	Marginal	Chief
Input	u_{i-1}	$\bar{y}_{i-1}^{\#}$
Output	$y_{i+1} = fu_{i-1}n_{i-1}$	$\bar{u}_{i+1} = -\bar{y}_{i-1}^{\#}/n_{i+1}f$

In Chap.7 on the fifth order, we will need to calculate, separately, the third-order contributions on each of the surfaces within a module. For the forward module, the intermediate quantities defined in (1.2.11) and discussed in (3.3.5) and (6.4.24) become

$$s_j = u_{j-1}n_{j-1}S_1 \quad , \quad \bar{s}_j = (\bar{y}_{j-1}^{\#}/f)\bar{S}_1 \quad ,$$
$$s_{j+1} = u_{j-1}n_{j-1}S_2 \quad , \quad \bar{s}_{j+1} = (\bar{y}_{j-1}^{\#}/f)\bar{S}_2 \quad , \tag{6.5.6}$$

where s_i and $\bar{S}_i$ $(i=j,j+1)$ are given in (6.4.24). From Table 6.2, we obtain the refraction invariants

$$a_j = u_{j-1}n_{j-1}A_1 \quad , \quad \bar{a}_j = (\bar{y}_{j-1}^{\#}/f)\bar{A}_1 \quad ,$$
$$a_{j+1} = u_{j-1}n_{j-1}A_2 \quad , \quad \bar{a}_{j+1} = (\bar{y}_{j-1}^{\#}/f)\bar{A}_2 \quad . \tag{6.5.7}$$

We also need an appropriate version of the Petzval contribution, which we obtain from (1.2.15):

$$P_j = P_1/f \quad , \quad P_1 = c_1(1-\mu_1)/N_0 \quad ,$$
$$P_{j+1} = P_2/f \quad , \quad P_2 = c_2(1-\mu_2)/N_1 \quad . \tag{6.5.8}$$

6.6 The Backward Orientation

The results of this section will, in a sense, be mirror images of the formulas
arrived at in the preceeding section. Here, we consider a module in a backward
orientation, beginning at the ith surface of an optical system, as shown in
Fig.6.2. Again, we find relations between the canonical parameters of the
module, its power parameter f, and the lens parameters. These are assembled
in Table 6.5, where the Petzval sum can also be found. Note that, because the
module is turned around, the signs of the curvatures are reversed.

   As in the case of the forward orientation, the backward orientation can be
perceived as a device to which an input is applied and from which an output
is extracted. The input for the marginal ray is the height of that ray, $y_i^{\#}$,
in the i-1 medium. Because the ray's slope here is zero, $y_i^{\#}$ is constant and
equals y_{i-1} and y_i. Because $Y_2 = 1$, we may write this as $y_i = y_i^{\#} Y_2$.

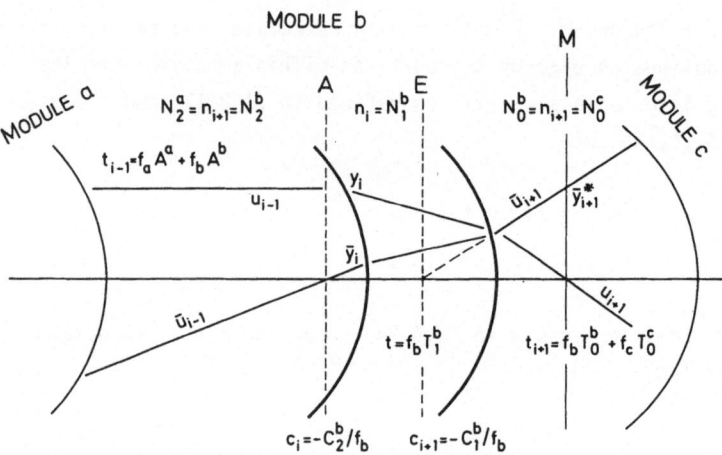

Fig. 6.2. The module as a component — the backward orientation

Table 6.5. Backward orientation — optical parameters

$n_{i-1} = N_2$	$c_i = -C_2/f$	$t_{i-1} = \ldots + fA$
$n_i = N_1$	$t_i = fT_1$	$t_{i+1} = fT_0 + \ldots$
$n_{i+1} = N_0$	$c_{i+1} = -C_1/f$	$= fE + \ldots$
$\bar{t} = fE$	$\tilde{t} = fG$	$\bar{t}' = fA$
$p = P/f$		

Table 6.6. Backward orientation — ray relationships

Marginal	Chief
$u_{i-1} = -(y_i^{\#}/f)U_2 = 0$	$\bar{u}_{i-1} = -n_{i-1}\bar{u}_{i-1}\bar{U}_2$
$y_i = y_i^{\#}Y_2$	$\bar{y}_i = fn_{i-1}\bar{u}_{i-1}\bar{Y}_2$
$a_i = -(y_i^{\#}/f)A_2$	$\bar{a}_i = -n_{i-1}\bar{u}_{i-1}\bar{A}_2$
$u_i = -(y_i^{\#}/f)U_1$	$\bar{u}_i = -n_{i-1}\bar{u}_{i-1}\bar{U}_1$
$y_{i+1} = y_i^{\#}Y_1$	$\bar{y}_{i+1} = fn_{i-1}\bar{u}_{i-1}\bar{Y}_1$
$a_{i+1} = -(y_i^{\#}/f)A_1$	$\bar{a}_{i+1} = -n_{i-1}\bar{u}_{i-1}\bar{A}_1$
$u_{i+1} = -(y_i^{\#}/f)U_0$	$\bar{u}_{i+1} = -n_{i-1}\bar{u}_{i-1}\bar{U}_0$
	$\bar{y}_{i+1}^{\#} = fn_{i-1}\bar{u}_{i-1}\bar{Y}_0$

$$h_i = -n_{i-1}\bar{u}_{i-1}y_i^{\#}H$$

We do the same sort of thing for the chief-ray input, which we take to be $\bar{u}_{i-1}$. From (6.3.4), $\bar{U}_2 = -1/N_2$, so that we may write $\bar{u}_{i-1} = -N_2\bar{u}_{i-1}\bar{U}_2 = (-n_{i-1}\bar{u}_{i-1})\bar{U}_2$. It follows that $\bar{y}_i = fn_{i-1}\bar{u}_{i-1}\bar{Y}_2$. By using the canonical ray-tracing equations (6.3.1-3), we can construct Table 6.6. Again, because the module has been turned around, we must be sure that the signs of the slopes of the rays are reversed.

The thing that is startlingly different about the backward module is the manner in which the input variables enter. In the case of the forward module, it was obvious that the variables $\bar{y}_0$, $\bar{u}_0$, and u_0 in (6.4.10) corresponded directly to the variables $\bar{y}_i$, $\bar{u}_{i-1}$, and u_{i-1}, respectively, in Table 6.2. However, in the case of the backward orientation, the variables in (6.4.10) correspond to the variables $\bar{y}_{i+1}^{\#}$, $\bar{u}_{i+1}$, and u_{i+1}, respectively, in Table 6.6. Therefore, when we calculate the aberrations for the backward orientation, we must be careful to insert these values rather than the input variables. We go through the same sequence of calculations that we used in the last section and obtain Table 6.7.

There remains the tabulation of the output quantities for the backward module. We have already encountered these, and need only collect them in Table 6.8.

As in the last section, we must include means for calculation of third-order contributions on the individual surfaces of the module. These will be required for the extrinsic fifth-order contributions. For the backward modules, the intermediate quantities of (1.2.11), (3.3.5), and (6.4.24) become

Table 6.7. Backward orientation — aberration contributions

Image coma	$\mathfrak{b}_i = n_{i-1}\bar{u}_{i-1}y_i^{\#3}F/f^2$
Pupil astigmatism	$\bar{c}_i = (n_{i-1}\bar{u}_{i-1})^2 y_i^{\#2}\bar{C}/f^2$
Pupil coma	$\bar{\mathfrak{b}}_i = -(n_{i-1}\bar{u}_{i-1})^3 y_i^{\#}\bar{F}$
Pupil spherical	$\mathfrak{b}_i = f(n_{i-1}\bar{u}_{i-1})^4\bar{B}$
Petzval	$p_i = P/f$
Image distortion	$e_i = -(n_{i-1}\bar{u}_{i-1})y_i^{\#}E$
Pupil distortion	$\bar{e}_i = -n_{i-1}\bar{u}_{i-1}y_i^{\#3}\bar{E}/f^2$
Longitudinal color	$\ell_i = y_i^{\#2}\mathscr{L}/f$
Transverse color	$t_i = -n_{i-1}\bar{u}_{i-1}y_i^{\#}\mathscr{T}$

Table 6.8. Backward orientation — module input and output

	Marginal	Chief
Input	$y_i^{\#}$	$\bar{u}_{i-1}$
Output	$u_{i+1} = -y_i^{\#}/fn_{i+1}$	$\bar{y}_{i+1}^{\#} = fn_{i-1}\bar{u}_{i-1}$

$$s_j = (y_j^{\#}/f)S_2 \quad, \quad \bar{s}_j = n_{j-1}\bar{u}_{j-1}\bar{S}_2 \quad,$$
$$s_{j+1} = (y_j^{\#}/f)S_1 \quad, \quad \bar{s}_{j+1} = n_{j-1}\bar{u}_{j-1}\bar{S}_1 \quad, \tag{6.6.1}$$

where s_i and $\bar{S}_i$ $(i=j,j+1)$ are found in (6.4.24). From Table 6.6, we get the refraction invariants

$$a_j = -(y_j^{\#}/f)A_2 \quad, \quad \bar{a}_j = -n_{j-1}\bar{u}_{j-1}\bar{A}_2 \quad,$$
$$a_{j+1} = -(y_j^{\#}/f)A_1 \quad, \quad \bar{a}_{j+1} = -n_{j-1}\bar{u}_{j-1}\bar{A}_1 \quad . \tag{6.6.2}$$

We will also require expressions for the Petzval contributions, which are identical to those for the forward module found in (6.5.8).

6.7 Hard-Way-Coupled Modules

In this section, we conjoin two modules in a hard-way coupling, as illustrated in Fig.6.3. We will indicate the canonical quantities associated with the first module by the subscript a, and those associated with the second module by the subscript b. Of course, the first module is in a backward orientation, whereas the second is in a forward orientation. Furthermore, we assume that, at least, a necessary condition for coupling is satisfied; that is, that the modules are in categories AB, BA, AD, DA, BC, or CB. The optical parameters are displayed in Table 6.9.

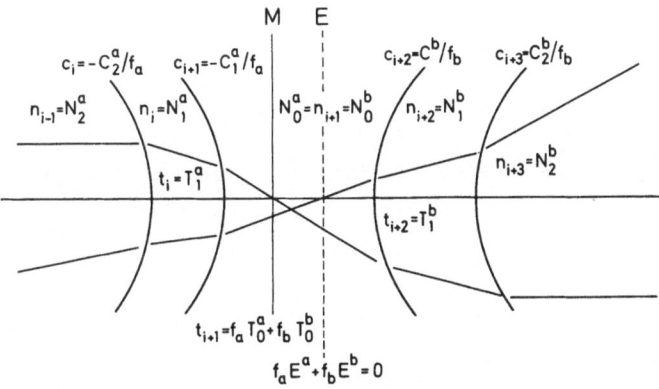

Fig. 6.3. The module as a component — a hard-way-coupled pair

Table 6.9. Hard-way coupled modules — optical parameters

$t_{i-1} = \ldots + f_a A^a$	$n_{i-1} = N_2^a$
$c_i = -C_2^a/f_a$	
$t_i = f_a T_1^a$	$n_i = N_1^a$
$c_{i+1} = -C_1^a/f_a$	
$t_{i+1} = f_a T_0^a + f_b T_0^b$	$n_{i+1} = N_0^a = N_0^b$
$c_{i+2} = C_1^b/f_b$	
$t_{i+2} = f_b T_1^b$	$n_{i+2} = N_1^b$
$c_{i+3} = C_2^b/f_b$	
$t_{i+3} = f_b A^b + \ldots$	$n_{i+3} = N_2^b$

In the last section, in Figs.5.8-10, we saw the $y-\bar{y}$ diagram for the hard-way coupling. As POWELL [6.4] has shown, the conditions for a successful coupling are particularly clear in this form.

We now have sufficient conditions that must be satisfied to assure a hard-way coupling. The first of these is that

$$f_a E^a + f_b E^b = 0 \quad ,$$

so that

$$f_b = -f_a E^a / E^b \quad . \tag{6.7.1}$$

In addition, it must be that

$$t_{i+1} = f_a (T_0^a - T_0^b E^a / E^b) > 0 \quad . \tag{6.7.2}$$

Finally, all thicknesses must be positive. In this case, to cover the case of reflecting surfaces, in which the index and the thickness are both negative, we require that

$$t_i n_i > 0 \quad , \quad t_{i+2} n_{i+2} > 0 \quad . \tag{6.7.3}$$

Table 6.10. Hard-way-coupled modules — ray variables

Marginal	Chief
$u_{i-1} = -(y_i^{\#}/f_a) U_2^a = 0$	$\bar{u}_{i-1} = -n_{i-1}\bar{u}_{i-1}\bar{U}_2^a$
$y_i = y_i^{\#}Y_2^a$	$\bar{y}_i = f_a n_{i-1}\bar{u}_{i-1}\bar{Y}_2^a$
$u_i = (y_i^{\#}/f_a) U_1^a$	$\bar{u}_i = -n_{i-1}\bar{u}_{i-1}\bar{U}_1^a$
$y_{i+1} = y_i^{\#}Y_1^a$	$\bar{y}_{i+1} = f_a n_{i-1}\bar{u}_{i-1}\bar{Y}_1^a$
$u_{i+1} = -(y_i^{\#}/f_a) U_0^a = -y_i^{\#}/f_a n_{i+1}$	$\bar{u}_{i+1} = -n_{i-1}\bar{u}_{i-1}\bar{U}_0^a$
	$\bar{y}_{i+1}^{\#} = f_a n_{i-1}\bar{u}_{i-1}\bar{Y}_0^a$
$y_{i+2} = u_{i+1}n_{i+1}Y_1^b = -(y_i/f_a)Y_1^b$	$\bar{y}_{i+2} = \bar{y}_{i+1}^{\#}Y_1^b = f_a n_{i-1}\bar{u}_{i-1}\bar{y}_{i+1}$
$u_{i+2} = f_b u_{i+1}n_{i+1}U_1^b = -(f_b y_i/f_a)U_1^b$	$\bar{u}_{i+2} = (\bar{y}_{i+1}^{\#}/f_b)\bar{U}_1^b = (f_a n_{i-1}\bar{u}_{i-1}/f_b)\bar{U}_1^b$
$y_{i+3} = u_{i+1}n_{i+1}Y_2^b = -(y_i^{\#}/f_a)Y_2^b$	$\bar{y}_{i+3} = \bar{y}_{i+1}^{\#}Y_2^b = f_a n_{i-1}\bar{u}_{i-1}Y_2^b$
$u_{i+3} = f_b u_{i+1}n_{i+1}U_2^b = 0$	$\bar{u}_{i+3} = (\bar{y}_{i+1}^{\#}/f_b)\bar{U}_2^b = (f_a n_{i-1}\bar{u}_{i-1}/f_b)\bar{U}_2^b$

Table 6.11. Hard-way-coupled modules — input and output

	Marginal	Chief
Input	$y_i^{\#}$	$\bar{u}_{i-1}$
Output	$y_{i+3} = -f_b y_i^{\#}/f_a$	$\bar{u}_{i+3} = -(f_a n_{i-1}/f_b n_{i+3})\bar{u}_{i-1}$

Table 6.12. Hard-way-coupled modules — aberration contributions

Image coma	$\displaystyle b = \left(y_i^{\#3} n_{i-1}\bar{u}_{i-1}/f_a^2\right)(F_a - F_b)$
Pupil astigmatism	$\displaystyle \bar{c} = \left(y_i^{\#2} n_{i-1}^2 \bar{u}_{i-1}^2/f_a^2\right)\left(\bar{C}_a - \bar{C}_b E^b/E^a\right)$
Pupil coma	$\displaystyle \bar{b} = -y_i^{\#} n_{i-1}^3 \bar{u}_{i-1}^3 \left(\bar{F}_a + \bar{F}_b E^{b2}/E^{a2}\right)$
Pupil spherical	$\displaystyle \bar{b} = f_a n_{i-1}^4 \bar{u}_{i-1}^{-4}\left(\bar{B}_a - \bar{B}_b E^{b3}/E^{a3}\right)$
Petzval	$\displaystyle p = (1/f_a)\left(P_a - P_b E^b/E^a\right)$
Image distortion	$\displaystyle e = -y_i^{\#} n_{i-1}^3 \bar{u}_{i-1}^3 \left(E_a + E_b E^{b2}/E^{a2}\right)$
Pupil distortion	$\displaystyle \bar{e} = -\left(y_i^{\#3} n_{i-1}\bar{u}_{i-1}/f_a^2\right)(\bar{E}_a - \bar{E}_b)$
Longitudinal color	$\displaystyle \ell = \left(y_i^{\#2}/f_a\right)\left(\mathscr{L}_a - \mathscr{L}_b E^b/E^a\right)$
Transverse color	$\displaystyle t = -y_i^{\#} n_{i-1}\bar{u}_{i-1}\left(\mathscr{T}_a + \mathscr{T}_b\right)$

In Table 6.10, we display the ray variables and their canonical counterparts, collected and collated from Tables 6.2,3.

The input and output relationships are summarized in Table 6.11.

Finally, we come to the aberration contributions. These are obtained by simply combining the aberration contributions for the backward orientation and the forward orientation, given in Tables 6.7 and 6.3, respectively. The input and output variables given in Tables 6.8 and 6.4 are also used, as is the coupling condition given in (6.7.1). These results are summarized in Table 6.12.

6.8 The Lens-Design Equations

We now have at our disposal a great deal of information on the module. We understand its properties as an input-output device in both its forward and backward orientations. We are able to write down its contributions to all the third-order aberrations — pupil as well as image — and the primary chromatic aberrations. We have also studied these same properties for a hard-way-coupled pair of modules. In short, we have accumulated all the material we require to assemble coupled systems of modules into optical designs with pre-scribed properties. Moreover, it is in our power to cause any of the third-order aberrations or primary chromatic aberrations to vanish.

To show how this is done, consider a hard-way-coupled pair of modules that is then easy-way coupled to a third module, as shown in Fig.6.4. Table 6.13 shows the optical parameters and the canonical parameters of the system. The

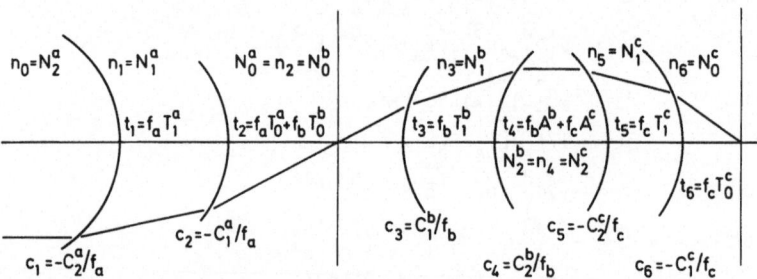

Fig. 6.4. A three-module system

Table 6.13. The three-module system — optical parameters

$$t_0 = A^a f_a \qquad n_0 = N_2^a$$

$$c_1 = -C_2^a/f_a$$

$$t_1 = T_1^a f_a \qquad n_1 = N_1^a$$

$$c_2 = -C_1^a/f_a$$

$$t_2 = T_0^a f_a + T_0^b f_b \qquad n_2 = N_0^a = N_0^b$$

$$c_3 = C_1^b/f_b$$

$$t_3 = T_1^b f_b \qquad n_3 = N_1^b$$

$$c_4 = C_2^b/f_b$$

$$t_4 = A^b f_b + A^c f_c \qquad n_4 = N_2^b = N_2^c$$

$$c_5 = -C_2^c/f_c$$

$$t_5 = T_1^c f_c \qquad n_5 = N_1^c$$

$$c_6 = -C_1^c/f_c$$

$$t_6 = T_0^c f_c \qquad n_6 = N_0^c$$

subscripts a, b, and c attached to the canonical quantities indicate that they belong to the first, second, or third module, respectively. t_0 is the distance from the entrance pupil to the first surface, and t_6 is the back focal distance. Possible stop positions are not indicated in this scheme.

If the third module is in category A or B, then the system forms a real image of an object at infinity at the main focus of the third module. On the other hand, if the third module is in category C or D, then the image is virtual and the system may be considered as a telescope for direct viewing. Moreover, if the device is in category D, the pupil is real and its distance to the last surface may be controlled to provide eye relief. These details will be considered further in Chap.8.

We assume that all conditions for both hard-way and easy-way coupling are satisfied, so that, from (6.7.1-3),

$$f_b = -f_a E^a / E^b \qquad (6.8.1)$$

$$t_2 = f_a (T_0^a - T_0^b E^a / E^b) > 0 \qquad (6.8.2)$$

$$t_1 n_1 > 0 \quad , \quad t_3 n_3 > 0 \quad , \quad t_5 n_5 > 0 \quad , \qquad (6.8.3)$$

as well as

$$t_4 = f_b A^b + f_c A^c > 0 \quad . \qquad (6.8.4)$$

From this point on, let us assume that object space and image space are both air, so that $n_0 = n_6 = 1$. Now, consider the input-output properties of the lens. We combine Table 6.11, concerning the hard-way-coupled modules, with Table 6.8 on the backward module and place the results in Table 6.14. Applying (6.8.1) to these results yields, for the fourth surface,

Table 6.14. The three-module system — input and output

	Marginal	Chief
Input	y_1	$\bar{u}_0$
Hard-way coupled Output Input	$y_4 = y_5 = -y_1 f_b / f_a$	$\bar{u}_4 = -\bar{u}_0 f_a / f_b n_4$
Backward Output	$u_6 = -y_4 / f_c = y_1 f_b / f_a f_c$	$\bar{y}_7^\# = f_c n_4 \bar{u}_4 = -\bar{u}_0 f_a f_c / f_b$

$$y_4 = y_1 E^a/E^b \quad , \quad \bar{u}_4 = \bar{u}_0 E^b/E^a n_4 \quad , \tag{6.8.5}$$

and for the image plane,

$$u_6 = -y_1 E^a/E^b f_c \quad , \quad \bar{y}_7^\# = \bar{u}_0 f_c E^b/E^a \quad . \tag{6.8.6}$$

Next, suppose that we wish the focal length of this lens to be $\mathscr{F}$. Then, because

$$\bar{y}_7^\# = \mathscr{F} \bar{u}_0 \quad ,$$

it follows that

$$f_c = \mathscr{F} E_a/E_b \quad . \tag{6.8.7}$$

Thus, the prescribed focal length of the lens determines f_c as a function of the shape parameters of the first two modules. We have already seen that f_b is determined, by the sufficient condition for a hard-way coupling, to be proportional to f_a and to depend also on the shape parameters of the first two modules.

Now let us prescribe an f number, defined for the case of the infinite object as the ratio of the focal length to the diameter of the entrance pupil. If $\mathscr{B}$ is the f number, then, in terms of the input quantities in Table 6.14,

$$\mathscr{B} = \mathscr{F}/2y_1 \quad ,$$

so that

$$y_1 = \mathscr{F}/2\mathscr{B} \quad . \tag{6.8.8}$$

Finally, suppose that the maximum semifield angle of the lens is to be β. Then we have

$$\bar{u}_0 = \tan\beta \quad . \tag{6.8.9}$$

In this way, all input quantities are determined.

At this stage, note that f_a and the three shape parameters associated with the three modules are undetermined. These can be ψ_a, ψ_b, ψ_c, or ϕ_a, ϕ_b, ϕ_c, or any combination thereof. It is not to be assumed that the range of these

parameters is unrestricted. Condition on coupling will confine these quantities to regions in which the module is in the proper category.

Following the pattern established in the earlier sections we next calculate the aberration contributions. We consider in detail image coma and add
together the expressions for this aberration from Table 6.12, the hard-way-
coupled modules, and Table 6.7, the backward module. We use input and output
variables from Table 6.14 and obtain

$$\oint = y_1^3 \bar{u}_0 n_0 (F_a - F_b)/f_a^2 + n_4 \bar{u}_4 y_5^3 F_c/f_c^2 \quad .$$

By setting $n_0 = 1$ and by substituting for $\bar{u}_4$ and y_5 from Table 6.14, we get

$$\oint = y_1^3 \bar{u}_0 \left[(F_a - F_b)/f_a^2 + F_c f_b^2/f_a^2 f_c^2 \right] \quad .$$

Finally, by eliminating f_b and f_c by means of (6.8.1,7), we obtain

$$\oint = y_1^3 \bar{u}_0 \left[(F_a - F_b)/f_a^2 + F_c / \mathscr{F}^2 \right] \quad . \tag{6.8.10}$$

This equation embodies the whole point of modular optical design. By setting $\oint$ equal to zero and solving for f_a, we obtain a value of the power parameter of the first module which, when substituted in the proper equations
in Table 6.13 and (6.8.1-9), results in an optical design in which third-
order spherical aberration, third-order coma, and third-order astigmatism are
identically zero. Moreover, within limits, all three shape parameters may be
altered, and f_a may be recalculated to obtain new systems with these same
properties. Thus the remaining nonzero aberrations, indeed, all other properties of the lens may be adjusted, within limits, without disturbing these
properties.

A procedure similar to that used to obtain (6.8.10) can be applied to the
other aberrations in the system. These have all been collected in Table 6.15.

By setting equal to zero any of the expressions for the aberrations given
in Table 6.15, we obtain an equation whose solution leads to an optical system that has the property that that particular aberration is identically zero.
Not all of these can be solved in terms of f_a, the first-power parameter.
Note that the equations for image coma, pupil astigmatism and pupil distortion are quadratic in $1/f_a^2$, whereas those for Petzval and longitudinal are
linear in $1/f_a$. Pupil spherical aberration, an oddball, depends on f_a. Pupil
coma, image distortion, and transverse color are independent of f_a and depend
only on the three shape parameters.

Table 6.15. The three-module system — aberration contributions

Image coma	$\delta = y_1^3\bar{u}_0\left[(F_a-F_b)/f_a^2+F_c/\mathcal{F}^2\right]$
Pupil astigmatism	$\bar{c} = y_1^2\bar{u}_0^{-2}\left[(\bar{C}_a-\bar{C}_bE^b/E^a)/f_a^2+\bar{C}_cE^{b2}/E^{a2}\mathcal{F}^2\right]$
Pupil coma	$\bar{\delta} = -y_1\bar{u}_0^{-3}\left[\bar{F}_a+(\bar{F}_b+\bar{F}_c)E^{b2}/E^{a2}\right]$
Pupil spherical	$\bar{b} = \bar{u}_0^{-4}\left[f_a(\bar{B}_a-\bar{B}_bE^{b3}/E^{a3})+\mathcal{F}\bar{B}_cE^{b3}/E^{a3}\right]$
Petzval	$p = (P_a-P_bE^b/E^a)/f_a + P_cE^b/E^a\mathcal{F}$
Image distortion	$e = -y_1\bar{u}_0^{-3}\left[E_a+(E_b+E_c)E^{b2}/E^{a2}\right]$
Pupil distortion	$\bar{e} = -y_1^3\bar{u}_0\left[(\bar{E}_a-\bar{E}_b)/f_a^2-\bar{E}_c/\mathcal{F}^2\right]$
Longitudinal color	$\ell = y_1^2\left[(\mathcal{L}_a-\mathcal{L}_bE^b/E^a)/f_a+\mathcal{L}_cE^a/E^b\mathcal{F}\right]$
Transverse color	$t = -y_1\bar{u}_0(\mathcal{F}_a+\mathcal{F}_b+\mathcal{F}_c)$

Table 6.16. The multimodule system — optical parameters. The system consists of $2n+1$ modules

$t_0 = f_1A_1$	$n_1 = N_{2,1}$
$c_{4\alpha-3} = -C_{2,2\alpha-1}/f_{2\alpha-1}$	
$t_{4\alpha-3} = f_{2\alpha-1}T_{1,2\alpha-1}$	$n_{4\alpha-3} = N_{1,2\alpha-1}$
$c_{4\alpha-2} = -C_{2,2\alpha-1}/f_{2\alpha-1}$	
$t_{4\alpha-2} = f_{2\alpha-1}T_{0,2\alpha-1} + f_{2\alpha}T_{0,2\alpha}$	$n_{4\alpha-2} = N_{0,2\alpha-1} = N_{0,2\alpha}$
$c_{4\alpha-1} = C_{1,2\alpha}/f_{2\alpha}$	
$t_{4\alpha-1} = f_{2\alpha}T_{1,2\alpha}$	$n_{4\alpha-1} = N_{1,2\alpha}$
$c_{4\alpha} = C_{2,2\alpha}/f_{2\alpha}$	
$t_{4\alpha} = f_{2\alpha}A_{2\alpha} + f_{2\alpha+1}A_{2\alpha+1}$	$n_{4\alpha} = N_{2,2\alpha}$
$t_{4n} = f_{2n}A_{2n} + f_{2n+1}A_{2n+1}$	
$c_{4n+1} = -C_{2,2n+1}/f_{2n+1}$	
$t_{4n+1} = f_{2n+1}T_{1,2n+1}$	
$c_{4n+2} = -C_{1,2n+1}/f_{2n+1}$	
$t_{4n+2} = f_{2n+1}T_{0,2n+1}$	

$= 1,2,\ldots,n$

This suggests a procedure for applying these equations. By solving first for the equations that are not dependent on the power parameter, we may find values of the shape parameters for which the corresponding aberrations vanish. Then, no matter what we do with the power parameters, these aberrations will always remain zero.

We have dwelt at great length on the three-module system, which is obviously only the simplest special case of a more-general arrangement. Consider a system consisting of n pairs of hard-way-coupled modules linked by an easy-way couple to a backward module, a total of 2n+1 modules. Such a system, like the three-module system just examined, will image an infinite object onto the main focus of the last module. As in the case of the three-module system, we can study this more general system by means of Tables 6.5-12. Table 6.16 gives the optical parameters. Here, we follow generally the notation originated by MERCADO [6.1]. In particular, we use Table 6.9 for the hard-way-coupled modules and Table 6.5 for the backward module.

Note the conditions for n hard-way-coupled modules, from (6.7.1):

$$f_{2\alpha-1}E_{2\alpha-1} + f_{2\alpha}E_{2\alpha} = 0 \quad , \quad \alpha = 1,2,\ldots,n \quad ,$$

so that

$$f_{2\alpha} = -f_{2\alpha-1}E_{2\alpha-1}/E_{2\alpha} \quad , \qquad \alpha = 1,2,\ldots,n \quad . \tag{6.8.11}$$

Now we turn to the input-output structure of the multimodule system described above. As in the three-module case, we take y_1 and $\bar{u}_0$, respectively, as the marginal-ray and chief-ray inputs to the system. Again, we make use of Table 6.11 for the hard-way-coupled modules and Table 6.8 for the last module, which is backward oriented.

Applying the results contained in Table 6.11 n times for the n pairs of hard-way-coupled modules, we obtain

$$y_{4n} = y_{4n+1} = (-1)^n y_1 f_{2n}f_{2n-1}\cdots f_2/f_{2n-1}f_{2n-3}\cdots f_1 \quad ,$$
$$n_{4n}\bar{u}_{4n} = (-1)^n n_0\bar{u}_0 f_{2n-1}f_{2n-3}\cdots f_1/f_{2n}f_{2n-2}\cdots f_2 \quad . \tag{6.8.12}$$

To the expressions in (6.8.12) we apply the results of Table 6.8 for the backward module and obtain

$$n_{4n+2}u_{4n+2} = \frac{(-1)^{n+1}y_1 f_{2n}f_{2n-2}\cdots f_2}{f_{2n+1}f_{2n-1}\cdots f}$$

$$\bar{y}^{\#}_{4n+2} = \frac{(-1)^n n_0 \bar{u}_0 f_{2n+1}f_{2n-1}\cdots f_1}{f_{2n}f_{2n-2}\cdots f_2} \quad . \tag{6.8.13}$$

Now, applying (6.8.11) repeatedly, we obtain

$$M = \frac{(-1)^n f_{2n}f_{2n-2}\cdots f_2}{f_{2n-1}\cdots f_1} = \frac{E_{2n-1}E_{2n-3}\cdots E_1}{E_{2n}E_{2n-2}\cdots E_2} \quad . \tag{6.8.14}$$

Note that M depends only on the shape parameters of the individual modules. Now, setting $n_0 = n_{4n+2} = 1$, we obtain from (6.8.13,14) the equations

$$u_{4n+2} = -My_1/f_{2n+1} \quad ,$$

$$\bar{y}^{\#}_{4n+2} = \bar{u}_0 f_{2n+1}/M \quad . \tag{6.8.15}$$

Finally, we let $\mathscr{F}$ represent the focal length of the system, $\mathscr{B}$ its f number, and β its semi-field angle. Then, as in (6.8.7-9),

$$f_{2n+1} = M\mathscr{F} \tag{6.8.16}$$

$$y_1 = \mathscr{F}/2\mathscr{B} \tag{6.8.17}$$

$$\bar{u}_0 = \tan\beta \quad . \tag{6.8.18}$$

Finally, we consider the aberration contributions that led to the lens-design equations. These are obtained in exactly the same way as the analogous expressions for the three-module system and are displayed in Table 6.17.

By setting the expressions for the aberration contributions given in Table 6.17 equal to zero, we obtain the lens-design equations. Note that they now may be regarded as simultaneous equations. Thus, in principle, by solving n of these equations simultaneously we obtain values for n power parameters, which leads to a system in which the corresponding n third-order aberrations vanish.

The three-module system, as well as the multimodule system treated in the foregoing, are lenses that image a point at infinity onto the main focus of the last module of the system. Other types are possible. The system described in Sect.6.7, consisting of two hard-way-coupled modules, is obviously an

Table 6.17. The multimodule system — aberration contributions

Image coma	$\delta = y_1^3 \bar{u}_0 \left[\sum_{\alpha=1}^{n} (F_{2\alpha-1} - F_{2\alpha})/f_{2\alpha-1}^2 + F_{2n+1}/\mathscr{F}^2 \right]$
Pupil astigmatism	$\bar{c} = y_1^2 \bar{u}_0^2 \left[\sum_{\alpha=1}^{n} (\bar{C}_{2\alpha-1} - \bar{C}_{2\alpha} E_{2\alpha}/E_{2\alpha-1})/f_{2\alpha-1}^2 + \bar{C}_{2n+1}/M^2 \mathscr{F}^2 \right]$
Pupil coma	$\bar{\delta} = -y_1 \bar{u}_0^3 \left[\sum_{\alpha=1}^{n} (\bar{F}_{2\alpha-1} + \bar{F}_{2\alpha} E_{2\alpha}^2/E_{2\alpha-1}^2) + \bar{F}_{2n+1}/M^2 \right]$
Pupil spherical	$\bar{b} = \bar{u}_0^4 \left[\sum_{\alpha=1}^{n} f_{2\alpha-1} (\bar{B}_{2\alpha-1} - \bar{B}_{2\alpha} E_{2\alpha}^3/E_{2\alpha-1}^3) + \mathscr{F} \bar{B}_{2n+1}/M^3 \right]$
Petzval	$p = \sum_{\alpha=1}^{n} (P_{2\alpha-1} - P_{2\alpha} E_{2\alpha}/E_{2\alpha-1})/f_{2\alpha-1} + P_{2n+1}/M\mathscr{F}$
Image distortion	$e = -y_1 \bar{u}_0^3 \left[\sum_{\alpha=1}^{n} (E_{2\alpha-1} + E_{2\alpha} E_{2\alpha}^2/E_{2\alpha-1}) + E_{2n+1}/M^2 \right]$
Pupil distortion	$\bar{e} = -y_1^3 \bar{u}_0 \left[\sum_{\alpha=1}^{n} (\bar{E}_{2\alpha-1} - \bar{E}_{2\alpha})/f_{2\alpha-1}^2 - \bar{E}_{2n+1}/\mathscr{F}^2 \right]$
Longitudinal color	$l = y_1^2 \left[\sum_{\alpha=1}^{n} (\mathscr{L}_{2\alpha-1} - \mathscr{L}_{2\alpha} E_{2\alpha}/E_{2\alpha-1})/f_{2\alpha-1} + \mathscr{L}_{2n+1} M/\mathscr{F} \right]$
Transverse color	$t = -y_1 \bar{u}_0 \left[\sum_{\alpha=1}^{n} (\mathscr{T}_{2\alpha-1} + \mathscr{T}_{2\alpha}) + \mathscr{T}_{2n+1} \right]$

afocal system with a magnification given by E_b/E_a (Table 6.11). By coupling
a number of these systems together, we obtain a multimodule afocal system,
as described in (6.8.12). Here, the magnification is given by M in (6.8.14).
Note that the magnification is independent of any power parameter and must
therefore be controlled by means of the shape parameter.

A finite-conjugate system is realized by easy-way coupling two modules,
the first in its forward orientation, the second in a backward orientation.
Because the coupling is an easy-way coupling, both power parameters are available for manipulation. Indeed, the magnification of this system turns out to
be the ratio of the two power parameters.

A multimodule finite-conjugate system retains this property. The magnification of the system depends on the power parameters of the first and the last
module. The others in the system are available for correcting aberrations.

A discussion of the details of such systems, the afocal and the finite-
conjugate system, will be postponed to Chap.8.

7. The Fifth Order

The concept of fifth-order aberrations proceeds logically and arithmetically from the idea of third-order aberrations. The latter are the third-order terms of a power-series expansion of an eikonal function. The fifth-order aberrations must then be identified with the fifth-order terms of the same power series. In principle, we can proceed to higher and higher orders of aberrations *ad infinitum*.

The order of difficulty in deriving, calculating and comprehending aberrations of the fifth-order far exceeds the mere increment 2 to the exponent. It is necessary to distinguish between two types of contributions to the fifth-order aberrations at a particular surface in a lens. The first type we will call *intrinsic*; they are associated with the terms of degree 5 in the power-series expansion of the eikonal at that surface. The other contributions, which, for want of a better term, we will call *extrinsic*, constitute fifth-order terms made up out of third-order coefficients from preceding surfaces of the lens. Therein lies the greatest complication that arises out of the use of the fifth-order quantities. Unlike the third order, fifth-order aberrations calculated at a given surface in an optical system depend on the past history of the paraxial rays used in its calculation. This is to say, the third-order calculations on the preceding surfaces enter into the calculation of the fifth-order aberrations.

The third-order aberrations were studied extensively by SCHWARZSCHILD [7.1] who devised what he called the Seidel eikonal, a function of five variables, whose first derivatives with respect to those variables yielded the five third-order aberration coefficients. KOHLSCHUTTER [7.2] is said to have calculated fifth-order aberration coefficients by taking second derivatives of the Seidel eikonal. The story goes that, because the formulas did not account for the extrinsic contributions, they are considered invalid. Perhaps another reason for their hasty abandonment is expressed by a comment by WACHENDORF [7.3] that "the formulas of Kohlschüter are so unmanageable that one must shrink in horror from them." WACHENDORF himself, as well as HERZBERGER [7.4], were among the more notable contributors to the lore of the fifth order. HERZBERGER's book contains a somewhat different historical treatment as well as references to his own work and the work of others on the fifth order.

The work presented here is based on BUCHDAHL [7.5]. The notation used is the form developed at the University of Rochester and recorded by RIMMER [7.6]. This work was carried out by ANDERSON [7.7,8] at the University of Arizona.

7.1 The Intrinsic Contributions

The nature of the fifth-order aberrations has been discussed more than adequately (some would say *ad nauseam*) elsewhere. Here we will be content to follow RIMMER [7.6] adapting his equations to the notations used in this work. Tables 7.1-3 present a comparison between his notation and ours, a set of

Table 7.1. Notation comparison — paraxial quantities

Rimmer's	Ours	Equations	Relation
n	n_{i+1}		$n = n_{j+1}$
			$n_{-1} = n_j$
$k = n_{-1}/n$	$\mu_{j+1} = n_j/n_{j+1}$	1.2.3	$k = \mu_{j+1}$
$i = yc + u_{-1}$	$a_{j+1} = n_j(c_{j+1}y_{j+1}+u_j)$	1.2.4	$i = a_{j+1}/n_j$
	$= n_{j+1}(c_{j+1}y_{j+1}u_{j+1})$		
$i' = ik$			$i' = a_{j+1}/n_{j+1}$
$u = u_{-1} + i(k-1)$	$u_{j+1} = \mu_{j+1}u_j$	1.2.1,2	$u = u_{j+1}$
	$- (1-\mu_{j+1})c_{j+1}y_{j+1}$		$u_{-1} = u_j$
$y_{+1} = y + tu$	$y_{j+1} = y_j + t_ju_j$	1.2.5-8	$y = y_j$
$I = n(y\bar{u}-u\bar{y})$	$h = n_j(\bar{u}_jy_j-u_j\bar{y}_j)$	1.2.9,10	$I = h$

Table 7.2. Notation comparison. Third-order quantities

Rimmer's	Ours	Equations	Relation
$s = n_{-1}(k-1)y(i+u)$	$s_{j+1} = -(1-\mu_{j+1})[(1+\mu_{j+1})u_j$	1.2.11	$S = y_{j+1}s_{j+1}N_j$
	$+\mu_{j+1}c_{j+1}y_{j+1}]/N_j$		
$B = Si^2$	$b_j = -a_j^2y_js_j$	1.2.12	$B = -b_{j+1}$
$F = Si\bar{\imath}$	$\not{b}_j = -a_j\bar{a}_jy_js_j$	1.2.13	$F = -\not{b}_{j+1}$
$C = S\bar{\imath}^2$	$c_j = -\bar{a}_j^2y_js_j$	1.2.14	$C = -c_{j+1}$
$\bar{B} = S\bar{\imath}^2$	$b_j = -\bar{a}_j^2\bar{y}_j\bar{s}_j$	1.2.24	$\bar{B} = -\bar{b}_{j+1}$
$\bar{F} = S\bar{\imath}i$	$\not{b}_j = -\bar{a}_ja_j\bar{y}_js_j$	1.2.25	$\bar{F} = -\not{\bar{b}}_{j+1}$
$\bar{C} = Si^2$	$c_j = -a_j^2\bar{y}_js_j$	1.2.26	$\bar{C} = -\bar{c}_{j+1}$
$E = \bar{F} + I(k-1)\bar{\imath}(\bar{u}+\bar{u}_{-1})$	$e_j = \not{b}_j - h(\bar{u}_j^2-\bar{u}_{j-1}^2)$	1.2.28	$E = -e_{j+1}$
$\bar{E} = F - I(k-1)i(u+u_{-1})$	$\bar{e}_j = \not{b}_j - h(u_j^2-u_{j-1}^2)$	1.2.23	$\bar{E} = -\bar{e}_{j+1}$
$\pi = \bar{\pi} = c(k-1)I^2/n_{-1}$	$p_j = c_j(1-\mu_j)/N_{j-1}$	1.2.15	$\pi = -p_{j+1}h^2$

Table 7.3. Notation comparison. Fifth-order quantities

Rimmer's	Ours
$\omega = (i^2+i'^2+u^2-3u_{-1}^2)/8$	$w_j = [a_j^2(1+u_j^2)/n_{j-1}^2+u_j^2-3u_{j-1}^2]/8$
$X_{73} = 3ii' + 2u^2 - 3u_{-1}^2$	$(xa)_j = 3a_j^2/n_jn_{j-1} + 2u_j^2 - 3u_{j-1}^2$
$X_{74} = 3i\bar{i}' + 2u\bar{u} - 3u_{-1}\bar{u}_{-1}$	$(xb)_j = 3a_j\bar{a}_j/n_jn_{j-1} + 2u_j\bar{u}_j - 3u_j\bar{u}_{j-1}$
$X_{75} = 3\bar{i}\bar{i}' + 2\bar{u}^2 - 3\bar{u}_{-1}^2$	$(xc)_j = 3\bar{a}_j^2/n_jn_{j-1} + 2\bar{u}_j^2 - 3\bar{u}_{j-1}^2$
$X_{76} = i(3u_{-1}-u)$	$(xe)_j = a_j(3u_{j-1}-u_j)/n_{j-1}$
$X_{77} = \bar{i}(2u_{-1}-u) + i\bar{u}_{-1}$	$(xf)_j = [\bar{a}_j(2u_{j-1}-u_j)+a_j\bar{u}_{j-1}]/n_{j-1}$
$X_{78} = \bar{i}(3\bar{u}_{-1}-\bar{u})$	$(xg)_j = \bar{a}_j(3\bar{u}_{j-1}-\bar{u}_j)/n_{j-1}$
$X_{42} = \bar{y}^2ci + y\bar{i}(\bar{u}+\bar{u}_{-1})$	$(xh)_j = [\bar{y}_j^2c_ja_j+y_j\bar{a}_j(\bar{u}_j+\bar{u}_{j-1})]/n_{j-1}$
$X_{82} = \bar{y}^2cu_{-1} - y\bar{i}'(\bar{u}+\bar{u}_{-1})$	$(xk)_j = \bar{y}_j^2c_ju_j - y_j\bar{a}_j(\bar{u}_j+\bar{u}_{j-1})/n_j$
$\bar{X}_{42} = y^2c\bar{i} + \bar{y}i(u+u_{-1})$	$(xm)_j = [y_j^2c_j\bar{a}_j+\bar{y}_ja_j(u_j+u_{j-1})]/n_{j-1}$
$\bar{X}_{82} = y^2c\bar{u}_{-1} - \bar{y}i'(u+u_{-1})$	$(xn)_j = y_j^2c_j\bar{u}_{j-1}-\bar{y}_ja_j(u_j+u_{j-1})/n_j$
$\hat{S}_{1p} = 3\omega Si/2$	$(sa)_j = 3w_jsy_ju_{j-1}a_j/2$
$\hat{S}_{2p} = S(\bar{i}X_{73}+iX_{74}-\bar{u}X_{76}-uX_{77})/4$	$(sb)_j = y_js_jn_{j-1}^2[(\bar{a}_j(xa)_j+a_j(xb)_j)/n_{j-1}$ $-(\bar{u}_j(xe)_j+u_j(xf)_j)]/4$
$\hat{S}_{3p} = n_{-1}(k-1)[X_{42}X_{73}+X_{76}X_{82}+y(i+u)$ $(iX_{75}-uX_{78})]/8$	$(sc)_j = n_{j-1}(u_j-1)\Big\{(xh)_j(xa)_j+(xe)_j(xk)_j+y_j$ $\times[a_j/n_j+u_j][a_j(xc)_j/n_{j-1}-u_j(xg)_j]\Big\}/8$
$\hat{S}_{4p} = S(\bar{i}X_{74}-\bar{u}X_{77})/2$	$(sd)_j = y_js_jn_{j-1}^2[\bar{a}_j(xb)_j/n_{j-1}-\bar{u}_j(xf)_j]/2$
$\hat{S}_{5p} = n_{-1}(k-1)[X_{42}X_{74}+X_{77}X_{82}+y(i+u)$ $(\bar{i}X_{75}-\bar{u}X_{78})]/4$	$(se)_j = n_{j-1}(u_j-1)\Big\{(xh)_j(xb)_j+(xf)_j(xk)_j$ $+y_j(a_j/n_{j-1}+u_j)[\bar{a}_j(xc)_j/n_{j-1}-\bar{u}_j(xg)_j]\Big\}/4$
$\hat{S}_{6p} = n_{-1}(k-1)(X_{42}X_{75}+X_{78}X_{82})/8$	$(sf)_j = n_{j-1}(u_j-1)[(xh)_j(xc)_j+(xg)_j(xk)_j]/8$
$\hat{S}_{1q} = n_{-1}(k-1)(\bar{X}_{42}X_{73}+X_{76}\bar{X}_{82})/8$	$(sg)_j = n_{j-1}(u_j-1)[(xm)_j(xa)_j+(xe)_j(xn)_j]/8$

Rosetta stones, to provide a suitable translation and correlation. Table 7.1 is concerned with the notations used for paraxial quantities and relates to (1.2.10). Table 7.2 involves third-order quantities and depends on (1.2.11-15). In Tables 7.3,4 the intrinsic fifth-order quantities are introduced and defined. To Table 7.4, we can add several important combinations: Tangential Oblique Spherical Aberration (TOBSA) at the jth surface of an optical system

$$TA_{T5} = (yd)_j + (ye)_j + (yf)_j \tag{7.1.1}$$

and Sagittal Oblique Spherical Aberration (SOBSA), by

$$TA_{S5} = (ye)_j \;\; ; \tag{7.1.2}$$

Table 7.4. Notation comparison. Intrinsic fifth-order aberrations. Quantities in parentheses are assumed to have a subscript $j+1$

Rimmer's	Ours
Fifth-Order Spherical Aberration	
$\tilde{B}_5 = 2i\hat{S}_{1p}$	$(ya)_j = 2a_j(sa)_j/n_{j-1}$
Fifth-Order Coma	
$\tilde{F}_1 = 2\bar{i}\hat{S}_{1p} + i\hat{S}_{2p}$	$(yb)_j = 2\bar{a}_j(sa)/n_{j-1} + (yc)_j$
$\tilde{F}_2 = i\hat{S}_{2p}$	$(yc)_j = a_j(sb)_j/n_{j-1}$
Oblique Spherical Aberration	
$\tilde{M}_1 = 2\bar{i}\hat{S}_{2p}$	$(yd)_j = 2\bar{a}_j(sb)_j/n_{j-1}$
$\tilde{M}_2 = 2i\hat{S}_{3p}$	$(ye)_j = 2a_j(sc)_j/n_{j-1}$
$\tilde{M}_3 = 2i\hat{S}_{4p}$	$(yf)_j = 2a_j(sd)_j/n_{j-1}$
Elliptical Coma	
$\tilde{N}_1 = 2\bar{i}\hat{S}_{3p}$	$(yg)_j = 2\bar{a}_j(sc)_j/n_{j-1}$
$\tilde{N}_2 = 2\bar{i}\hat{S}_{4p} + 2i\hat{S}_{5p}$	$(yh)_j = 2\bar{a}_j(sd)_j/n_{j-1} + 2(yk)_j$
$\tilde{N}_3 = i\hat{S}_{5p}$	$(yk)_j = a_j(se)_j/n_{j-1}$
Astigmatism	
$\tilde{C}_5 = \bar{i}\hat{S}_{5p}/2$	$(ym)_j = \bar{a}_j(se)_j/2n_{j-1}$
Fifth-Order "Petzval"	
$\tilde{\pi}_5 = 2i\hat{S}_{6p} - \bar{i}\hat{S}_{5p}/2$	$(yn)_j = 2a_j(sf)_j/n_{j-1} - (ym)_j$
Fifth-Order Image Distortion	
$\tilde{E}_5 = 2\bar{i}\hat{S}_{6p}$	$(yp)_j = 2\bar{a}_j(sf)_j/n_{j-1}$
Fifth-Order Pupil Distortion	
$\tilde{\bar{E}}_5 = 2i\hat{S}_{1q}$	$(yq)_j = 2a_j(sg)_j/n_{j-1}$

Elliptical Coma (ELLCOMA) is properly

$$LCOM_5 = (yg)_j + (yh)_j \quad ; \tag{7.1.3}$$

Transverse astigmatism and sagittal astigmatism are, respectively,

$$TAS_5 = (yn)_j + 5(ym)_j \tag{7.1.4}$$

and

$$SAS_5 = (yn)_j + (ym)_j \ .$$
(7.1.5)

In all of these definitions, a factor of 1/2h is omitted.

7.2 The Extrinsic Contributions

The extrinsic fifth-order contributions at a surface in an optical system
depend upon third-order quantities calculated for that surface and upon sums
of all third-order quantities calculated on preceding surfaces. In this section, we present these calculations. Table 7.5 gives RIMMER's [7.6] formulas
as well as the notation used here. Note that RIMMER used a prime (') to denote summation over all preceding surfaces. We prefer the more conventional
summation sign ($\sum$) which is a little less cryptic but a little more clumsy.
Note also that a factor 1/2h is omitted.

The application of these quantities to modular design results in both a
complication and a simplification. The extrinsic fifth-order quantities calculated at a given surface depend on sums of third-order quantities summed
over all preceding surfaces. Consider a modular design. The extrinsic fifth-
order contributions at the *first* surface of a module will depend on the sums
of third-order contributions over all preceding *modules*. The extrinsic fifth-
order contributions at the second surface of a module will depend on these
same third-order sums over the preceding modules plus contributions due to
the third-order contributions of the modules' first surface. Thus the total
extrinsic fifth-order contribution due to a given module consist of two parts:
the contribution due to the first surface in the module, and the contribution
due to *twice* the sum of the third-order quantities over the preceeding modules.
Thus does the fifth order accord the earlier modules in an optical system a
disproportionate influence over the quality of the overall design.

The modular approach does result in some simplification because the sums,
over a single module, of both third-order spherical aberration and astigmatism
are identically zero. Thus, all extrinsic contributions summed over complete
modules, with factors $\sum b_k$ and $\sum c_k$, are zero.

Table 7.5. Notation comparison. Fifth-order extrinsic quantities

Fifth-order spherical aberration

$$^OB_5 = 3(B'F-E'B)/2I$$

$$(za)_j = 3[\delta_{j+1} \sum^{j-1} b_k - b_{j+1} \sum^{j-1} \bar{e}_k]/2h$$

Fifth-order coma

$$^OF_1 = [B'(\pi+4c)+(5F'-4E')F-(2\pi'+5\bar{c}')B]/2I$$

$$(zb)_j = [(p_jh^2+4c_{j+1}) \sum^{j-1} b_k + \delta_j \sum^{j-1} (5\delta_k-4\bar{e}_k) - b_{j+1} \sum^{j-1} (2p_kh^2+5\bar{c}_k)]/2h$$

$$^OF_2 = [B'(\pi+2c)+2(2F'-E')F-(\pi'+4\bar{c}')B]/2I$$

$$(zc)_j = [(p_jh^2-2c_{j+1}) \sum^{j-1} b_k + 2\delta_{j+1} \sum^{j-1} (2\delta_k-\bar{e}_k)-b_{j+1} \sum^{j-1} (p_kh^2+4\bar{c}_k)]/2h$$

Oblique spherical aberration

$$^OM_1 = [B'E+(4F'-\bar{E}')c+(c'-4\bar{c}'-2\pi')F-F\bar{F}'B]/I$$

$$(zd)_j = [e_j \sum^{j-1} b_k+c_j \sum^{j-1} (4\delta_k-\bar{e}_k)+\delta_j \sum^{j-1} (c_k-4\bar{c}_k-2p_kh^2)-b_{j+1} \sum^{j-1} \bar{\delta}_k]/h$$

$$^OM_2 = [B'E+(2F'-\bar{E}')(\pi+c)+(\pi'+3c'-2\bar{c}')F-3F'B]/2I$$

$$(ze)_j = [e_j \sum^{j-1} b_k+(p_jh^2+c_j) \sum^{j-1} (2\delta_k-\bar{e}_k)+\delta_j \sum^{j-1} (p_kh^2+3c_k-2\bar{c}_k)-3b_j \sum^{j-1} \bar{\delta}_k]/2h$$

$$^OM_3 = 2[F'(2c+\pi)+(c'-2\bar{c}')F-F'B]/I$$

$$(zf)_j = 2[(2c_j+p_jh^2) \sum^{j-1} b_k+\delta_j \sum^{j-1} (c_k-2\bar{c}_k)-b_j \sum^{j-1} \bar{\delta}_k]/h$$

Astigmatism

$$^OC_5 = (4c'+\pi')E - \bar{F}'\pi + 2(E'-2F')C - 2\bar{B}'F/4I$$

$$(zm)_j = [e_j \sum^{j-1} (4c_k+p_kh^2+p_jh^2) - \sum^{j-1} \bar{\delta}_k+2c_j \sum^{j-1} (e_k-2\bar{\delta}_k)-2\delta_j \sum^{j-1} \bar{b}_k]/4h$$

Fifth-order "Petzval"

$$^O\pi_5 = [(\pi'-2c')E+(4E'-\bar{F}')\pi+2(\bar{E}'+\bar{F}')c-2\bar{B}'F]/4I$$

$$(zn)_j = [e_j \sum^{j-1} (p_kh^2-2c_k)+p_jh^2 \sum^{j-1} (4e_k-\bar{\delta}_k)+2c \sum^{j-1} (e_k+\bar{\delta}_k)-2\delta_j \sum^{j-1} \bar{b}_k]/4h$$

Elliptical coma

$$^ON_1 = [3F'E-(\pi'+\bar{c}')(\pi+c)+2(c'-\bar{c}')c+(E'-2F')F-\bar{B}'B]/2I$$

$$(zg)_j = [3e_j \sum^{j-1} \delta_k-(p_jh^2+c_j) \sum^{j-1} (p_kh^2+c_k)+2c_j \sum^{j-1} (c_k-\bar{c}_k)$$
$$+\delta_j \sum^{j-1} (e_k-2\bar{\delta}_k)-b_{j+1} \sum^{j-1} \bar{\delta}_k]/2h$$

$$^ON_2 = 2[3F'E+(3c'-\bar{c}'+\pi')(\pi+3c)-(\pi'+c')c+(E'-8F')F-\bar{B}'B]/I$$

$$(zh)_j = [3e_j \sum^{j-1} \delta_k+(p_jh^2+3c_j) \sum^{j-1} (3c_j-\bar{c}_k+p_kh^2)$$
$$-c_j \sum^{j-1} (p_kh^2+c_k)+\delta_j \sum^{j-1} (e_k-8\bar{\delta}_k)-b_j \sum^{j-1} \bar{\delta}_k]/h$$

$$^ON_3 = [F'E+(3c'+\pi'-\bar{c}')(\pi+c)+(c'+\pi')c+(E'-4F')F-\bar{B}'B]/2I$$

$$(zk)_j = [e_j \sum^{j-1} \delta_k+(p_jh^2+c_j) \sum^{j-1} (3c_j+p_k-\bar{c}_k)$$
$$+c_j \sum^{j-1} (c_k+p_kh^2+\delta_j) \sum^{j-1} (e_k-4\bar{\delta}_k)-b_j \sum^{j-1} \bar{\delta}_k]/2h$$

Fifth-order image distortion

$$^OE_5 = [3E'E-\bar{B}'(\pi+3c)]/2I$$

$$(zp)_j = [3e_{j+1} \sum^{j-1} e_k-(p_jh^2+3c_j) \sum^{j-1} \bar{\delta}_k]/2h$$

Fifth-order pupil distortion

$$^O\bar{E}_5 = [B'(\pi+3\bar{c})-3E'E]/2I$$

$$(zq)_j = [(p_jh^2+3\bar{c}_j) \sum^{j-1} b_k-\bar{e}_j \sum^{j-1} \bar{e}_k]/2h$$

7.3 The Forward Orientation

In this section, we assemble the formulas for the canonical versions of the
fifth-order aberrations over a module. These will be the intrinsic contributions from the formulas in Tables 7.3,4 and the extrinsic contributions in
Table 7.5. The subject is most complicated. Perhaps the best way to proceed
is to do some specific calculations. We follow the procedures laid down by
ANDERSON [7.7,8].

   We begin with the calculation of fifth-order spherical aberration for the
k^{th} module in an optical system. We assume in this section that the module is
in its forward orientation. The first surface of the kth module in an optical
system will have the index 2k-1, whereas the second will be labeled 2k.

   From Table 7.4, we find that the expression for the intrinsic fifth-order
spherical surface at the ith surface of an optical system is

$$(ya)_j = 2a_j(sa)_j/n_{j-1} \quad . \tag{7.3.1}$$

From Table 7.3 we find that

$$(sa)_j = 3w_j s_j y_j \mu_{j-1} a_j/2 \tag{7.3.2}$$

$$w_j = \left[a_j^2(1-\mu_j^2)/n_{j-1}^2 + u_j^2 - 3u_{j-1}^2 \right]/8 \quad . \tag{7.3.3}$$

We next introduce the idea of the canonical fifth-order contributions. Recall that the canonical terms of this order must be tagged to indicate whether
the orientation is forward or backward. To distinguish the two we use a superior + to denote the forward orientation and a - to signal the backward case.

   We begin with the last of the above equations, namely (7.3.3). We use expressions from (6.4.24) and Table 6.2 to obtain, for the first surface of the
k^{th} module,

$$
\begin{aligned}
W_{2k-1} &= \left[a_{2k-1}^2(1-\mu_{2k-1}^2)/n_{2k-1}^2 + u_{2k-1}^2 - 3u_{2k-2}^2 \right]/8 \\
&= u_{2k-2}^2 n_{2k-2}^2 \left[(A_1/N_0)^2(1-\mu_1)^2 + U_1^2 - 3U_0^2 \right]/8 \\
&= u_{2k-2}^2 n_{2k-2}^2 W_1^+ \quad ,
\end{aligned}
\tag{7.3.4}
$$

where

$$W_1^+ = \left[(A_1/N_0)^2(1-\mu_1)^2 + U_1^2 - 3U_0^2 \right]/8 \quad . \tag{7.3.5}$$

In exactly the same way we find that, for the modules' second surface

$$W_{2k} = u_{2k-2}^2 n_{2k-2}^2 W_2^+ \quad ,$$

where W_2^+ is identical to W_1^+ except that all of its indices are increased by unity.

Next, we treat (7.3.2) and obtain

$$
\begin{aligned}
(sa)_{2k-1} &= (3/2)w_{2k-1}s_{2k-1}y_{2k-1}\mu_{2k-2}a_{2k-1} \\
&= (3/2)(u_{2k-2}^2 n_{2k-2}^2 W_1^+(u_{2k-2}n_{2k-2}S_1)(f_k u_{2k-2}n_{2k-2}Y_1)\mu_1(u_{2k-2}n_{2k-2}A_1) \\
&= u_{2k-2}^5 n_{2k-2}^5 f_k(SA)_1^+ \quad ,
\end{aligned}
$$

where

$$(SA)_1^+ = 3W_1^+ S_1 Y_1 \mu_1 A_1/2 \quad . \tag{7.3.6}$$

A similar expression is obtained for this quantity on the second surface.

So far we have obtained canonical versions for two of the expressions that appear in Table 7.3 for the forward module. Table 7.6 contains the expressions for the rest of them.

Finally, we take up (7.3.1), the proper expression for fifth-order spherical aberration

$$
\begin{aligned}
(ya)_{2k-1} &= 2a_{2k-1}(sa)_{2k-1}/n_{2k-2} \\
&= 2[u_{2k-2}n_{2k-2}(A_1/N_0)][u_{2k-2}^5 n_{2k-2}^5 f_k(SA)_1^+] \\
&= u_{2k-2}^6 n_{2k-2}^6 f_k(YA)_1^+ \quad ,
\end{aligned}
\tag{7.3.7}
$$

where

$$(YA)_1^+ = 2(A_1/N_0)(SA)_1^+ \quad . \tag{7.3.8}$$

Equation (7.3.8) is now the expression for fifth-order spherical aberration for the first surface of the kth module. Thus the intrinsic fifth-order contribution for the entire module is given by

$$u_{2k-2}^6 n_{2k-2}^6 f_k[(YA)_1^+ + (YA)_2^+] \quad .$$

Table 7.6. Canonical intrinsic fifth-order aberrations

First surface only	Forward module
$W_{2k-1} = u_{2k-2}^2 n_{2k-2}^2 W_1^+$	$W_1^+ = (A_1/N_0)^2(1-u_1)^2 + U_1^2 - 3U_0^2$
$(xa)_{2k-1} = u_{2k-2}^2 n_{2k-2}^2 (XA)_1^+$	$(XA)_1^+ = 3A_1^2 N_1 + 2U_1^2 - 3U_0^2$
$(xb)_{2k-1} = (u_{2k-2}^2 n_{2k-2}^2 \bar{v}^2 {}_{2k-2}/f_k)(XB)_1^+$	$(XB)_1^+ = A_1\bar{A}_1/N_1 N_0 + 2U_1\bar{U}_1 - 3U_0\bar{U}_0$
$(xc)_{2k-1} = (\bar{y}^{\#2}_{2k-2}/f_k^2)(XC)_1^+$	$(XC)_1^+ = 3\bar{A}_1^2/N_1 N_0 + 2\bar{U}_1^2 - 3\bar{U}_0^2$
$(xe)_{2k-1} = u_{2k-2}^2 n_{2k-2}^2 (XE)_1^+$	$(XE)_1^+ = A_1(3U_0 - U_1)/N_0$
$(xf)_{2k-1} = (u_{2k-2}^2 n_{2k-2}^2 \bar{y}^{\#}_{2k-2}/f_k)(XF)_1^+$	$(XF)_1^+ = [\bar{A}_1(2U_0-U_1)+A_1\bar{U}_0]/N_0$
$(xg)_{2k-1} = (\bar{y}^{\#2}_{2k-2}/f_k^2)(XG)_1^+$	$(XG)_1^+ = \bar{A}_1(3\bar{U}_0 - \bar{U}_1)/N_0$
$(xh)_{2k-1} = (u_{2k-2}^2 n_{2k-2}^2 \bar{y}^{\#}_{2k-2}/f_k)(XH)_1^+$	$(XH)_1^+ = [\bar{V}_1^2 C_1 A_1 + Y_1\bar{A}_1(U_1+U_0)]/N_0$
$(xk)_{2k-1} = (u_{2k-2}^2 n_{2k-2}^2 \bar{y}^{\#2}_{2k-2}/f_k)(XK)_1^+$	$(XK)_1^+ = [\bar{V}_1^2 C_1 \bar{U}_0 - Y_1\bar{A}_1(U_1+U_0)]/N_0$
$(xm)_{2k-1} = u_{2k-2}^2 n_{2k-2}^2 (XM)_1^+$	$(XM)_1^+ = [\bar{V}_1^2 C_1 \bar{A}_1 + \bar{P}_1 A_1(U_1+U_0)]/N_0$
$(xn)_{2k-1} = u_{2k-2}^2 n_{2k-2}^2 (XN)_1^+$	$(XN)_1^+ = Y_1^2 C_1 \bar{U}_0 - \bar{P}_1 A_1(U_1+U_0)/N_1$
$(sa)_{2k-1} = u_{2k-2}^5 {}_{2k-2} f_k (SA)_1^+$	$(SA)_1^+ = 3W_1^5 Y_1 \mu_1 A_1/2$
$(sb)_{2k-1} = u_{2k-2}^4 n_{2k-2}^4 \bar{v}^{\#}_{2k-2}(SB)_1^+$	$(SB)_1^+ = Y_1 S N_1^2\{[\bar{A}_1(XA)_1^+ + A_1(XB)_1^+]/N_1 - [\bar{U}_1(XE)_1^+ + U_1(XF)_1^+]\}/4$
$(sc)_{2k-1} = (u_{2k-2}^3 n_{2k-2}^3 \bar{v}^{\#2}_{2k-2}/f_k)(SC)_1^+$	$(SC)_1^+ = N_0(u_1-1)\{(XH)_1^+(XA)_1^+ + (XE)_1^+(XK)_1^+ + Y_1(A_1/N_0+U_1)[(XC)_1^+/N_0-U_1(XG)_1^+]\}1/8$
$(sd)_{2k-1} = (u_{2k-2}^3 n_{2k-2}^3 \bar{v}^{\#2}_{2k-2}/f_k)(SD)_1^+$	$(SD)_1^+ = Y_1 S N_0^2 \bar{A}_1(XB)_1^+/N_0 - \bar{U}_1(XF)_1^+\}1/2$
$(se)_{2k-1} = (u_{2k-2}^2 n_{2k-2}^2 \bar{v}^{\#3}_{2k-2}/f_k^2)(SE)_1^+$	$(SE)_1^+ = N_0(u_1-1)\{(XH)_1^+(XB)_1^+ + (XF)_1^+(XK)_1^+ + Y_1(A_1/N_0+U_1)[\bar{A}_1(XC)_1^+/N_0-\bar{U}_1(XG)_1^+]\}1/4$
$(sf)_{2k-1} = (u_{2k-2}^2 n_{2k-2}^2 \bar{v}^{\#4}_{2k-2}/f_k^3)(SF)_1^+$	$(SF)_1^+ = N_0(u_1-1)[(XH)_1^+(XC)_1^+ + (XG)_1^+]1/8$
$(sg)_{2k-1} = (u_{2k-2}^4 n_{2k-2}^4 \bar{v}^{\#}_{2k-2})(SG)_1^+$	$(SG)_1^+ = N_0(u_1-1)[(XM)_1^+(XA)_1^+ + (XE)_1^+(XN)_1^+ + (XG)_1^+]1/8$

Table 7.7. Canonical intrinsic fifth-order aberrations—forward module

Fifth-Order Spherical Aberration

$(ya)_{2k-1} = u_{2k-2}^6 n_{2k-2}^6 f_k (YA)_1^+$ | $(YA)_1^+ = 2A_1(SA)_1^+/N_0$

Fifth-Order Coma

$(yb)_{2k-1} = u_{2k-2}^5 n_{2k-2}^5 \bar{y}^{\#}_{2k-2}(YB)_1^+$ | $(YB)_1^+ = 2\bar{A}_1(SA)_1^+/N_0 + (YC)_1^+$

$(yc)_{2k-1} = u_{2k-2}^5 n_{2k-2}^5 \bar{y}^{\#}_{2k-2}(YC)_1^+$ | $(YC)_1^+ = A_1(SB)_1^+/N_0$

Oblique Spherical Aberration

$(yd)_{2k-1} = (u_{2k-2}^4 n_{2k-2}^4 \bar{y}^{\#2}_{2k-2}/f_k)(YD)_1^+$ | $(YD)_1^+ = 2\bar{A}_1(SB)_1^+/N_0$

$(ye)_{2k-1} = (u_{2k-2}^4 n_{2k-2}^4 \bar{y}^{\#2}_{2k-2}/f_k)(YE)_1^+$ | $(YE)_1^+ = 2A_1(SC)_1^+/N_0$

$(yf)_{2k-1} = (u_{2k-2}^4 n_{2k-2}^4 \bar{y}^{\#2}_{2k-2}/f_k)(YF)_1^+$ | $(YF)_1^+ = 2A(SD)_1^+/N_0$

Elliptical Coma

$(yg)_{2k-1} = (u_{2k-2}^3 n_{2k-2}^3 \bar{y}^{\#3}_{2k-2}/f_k^2)(YG)_1^+$ | $(YG)_1^+ = 2\bar{A}_1(SC)_1^+/N_0$

$(yh)_{2k-1} = (u_{2k-2}^3 n_{2k-2}^3 \bar{y}^{\#3}_{2k-2}/f_k^2)(YH)_1^+$ | $(YH)_1^+ = 2\bar{A}_1(SD)_1^+/N_0 + 2(YK)_1^+$

$(yk)_{2k-1} = (u_{2k-2}^3 n_{2k-2}^3 \bar{y}^{\#3}_{2k-2}/f_k^2)(YK)_1^+$ | $(YK)_1^+ = A_1(SE)_1^+/N_0$

Fifth-Order Astigmatism

$(ym)_{2k-1} = (u_{2k-2}^2 n_{2k-2}^2 \bar{y}^{\#4}_{2k-2}/f_k^3)(YM)_1^+$ | $(YM)_1^+ = \bar{A}_1(SE)_1^+/N_0$

Fifth-Order "Petzval"

$(yn)_{2k-1} = (u_{2k-2}^2 n_{2k-2}^2 \bar{y}^{\#4}_{2k-2}/f_k^3)(YN)_1^+$ | $(YN)_1^+ = 2A_1(SF)_1^+/N_0 - (YM)_1^+$

Fifth-Order Image Distortion

$(yp)_{2k-1} = (u_{2k-2}^2 n_{2k-2}^2 \bar{y}^{\#5}_{2k-1}/f_k^4)(YP)_1^+$ | $(YP)_1^+ = 2\bar{A}_1(SF)_1^+/N_0$

Fifth-Order Pupil Distortion

$(yq)_{2k-1} = u_{2k-2}^5 n_{2k-2}^5 \bar{y}^{\#}_{2k-2}(YQ)_1^+$ | $(YQ)_1^+ = 2A(SG)_1^+/N_0$

We can construct, in exactly this manner, all of the fifth-order intrinsic aberration contributions for the forward module found in Table 7.4. These are found in Table 7.7. The canonical quantities shown in this table are for the first surface only. Those for the second surface are obtained by increasing each subscript by one.

Now we turn to the extrinsic fifth-order aberrations. It is convenient to refine those quantities even further. At a particular surface in an optical system, the extrinsic fifth-order contribution arises from the third-order contributions from all of the preceding surfaces. Thus, the extrinsic contributions associated with the first surface of a module will depend on third-order sums over all of the preceding modules. However, at the second surface of a module, the extrinsic contributions include contributions from that module's first surface as well as sums over all preceding modules. Thus we can consider two components of the extrinsic contribution; the extrinsic *internal* contribution due to third-order contributions on the module's first surface, and the extrinsic *external* contribution arising from the third-order quantities originating from each of the preceeding modules.

We again turn to the calculation of fifth-order spherical aberration. Refer now to Table 7.5,

$$(za)_{2k-1} + (za)_{2k} = 3\left[\ell_{2k-1} \sum^{2k-2} b_m - b_{2k-1} \sum^{2k-2} \bar{e}_m \right] / 2h$$

$$+ \left[3\ell_{2k} \sum^{2k-1} b_m - b_{2k} \sum^{2k-1} \bar{e}_m \right] / 2h$$

$$= 3\left\{ [\ell_{2k} b_{2k-1} - b_{2k} \bar{e}_{2k-1}] \right.$$

$$\left. + \left[(\ell_{2k-1} + \ell_{2k}) \sum^{2k-2} b_m - (b_{2k-1} + b_{2k}) \sum^{2k-2} \bar{e}_m \right] \right\} / 2h \quad .$$

The first pair of brackets encloses the extrinsic internal contribution, which we evaluate next,

$$[(za)_{2k-1} + (za)_{2k}]^{in} = 3[\ell_{2k} b_{2k-1} - b_{2k} \bar{e}_{2k-1}]/2h$$

$$= 3\left[(u_{2k-2}^3 n_{2k-2}^3 \bar{y}_{2k-2}^\# - 2F_2)(u_{2k-2}^4 n_{2k-2}^4 - 2f_k B_1) \right.$$

$$\left. - (u_{2k-2}^4 n_{2k-2}^4 - 2f_k B_2)(u_{2k-2}^3 n_{2k-2}^3 \bar{y}_{2k-2}^\# - 2\bar{E}_1) \right]$$

$$/2(-u_{2k-2} n_{2k-2} \bar{y}_{2k-2}^\# - 2H)$$

$$= u_{2k-2}^6 n_{2k-2}^6 - 2f_k (ZA)_{in}^+ \quad ,$$

where

$$(ZA)^+_{in} = -3(F_2B_1-B_2\bar{E}_1)/2H \quad .$$

All of the extrinsic internal contributions are collected in Table 7.8. The extrinsic external contribution comes next,

$$
\begin{aligned}
[(za)_{2k-1}+(za)_{2k}]^{ex} &= 3\left[(b_{2k-1}+b_{2k})\sum^{2k-2} b_m+(b_{2k-1}+b_{2k})\sum^{2k-2}\bar{e}_m\right]/2 \\
&= 3\left[u^3_{2k-2}n^3_{2k-2}\bar{y}^{\#}_{2k-2}(F_1+F_2)_k \sum^{k-1} u^4_{2m}n^4_{2m}f_m(B_1+B_2)_m \right. \\
&\quad \left. + u^4_{2k-2}n^4_{2k-2}f_k(B_1-B_2)_k \sum^{k-1} u^3_{2m}n^3_{2m}\bar{y}^{\#}_{2m}(F_1+F_2)_m\right] \\
&\quad /2(-u_{2k-2}n_{2k-2}\bar{y}^{\#}_{2k-2}H) \\
&= -3u^2_{2k-2}n^2_{2k-2}F_k \sum^{k-1} u^4_{2m}n^4_{2m}f_mB_m/2H \\
&\quad - 3(u^3_{2k-2}n^3_{2k-2}f_k/\bar{y}^{\#}_{2k-2})B_k \sum u^3_{2m}n^3_{2m}\bar{y}^{\#}_{2m}F_m/2H \quad .
\end{aligned}
$$

Here we use the third-order notation introduced in (6.4.13-19). The extrinsic external terms, of which these are typical, are subject to a further simplification, which will be treated subsequently. Of course, these extrinsic external contributions to fifth-order spherical aberration are all zero. This is because

$$\sum^{2k-2} b_m = 0 \qquad \text{and} \qquad b_{2k-1} + b_{2k} = 0 \quad .$$

In the other calculations that lead to the entries in Table 7.9, these equations, as well as $\sum^{2k-2} c_m = 0$ and $c_{2k-1} + c_{2k} = 0$, will be used to simplify things.

Table 7.8. Extrinsic Internal Contributions

Fifth-Order Spherical Aberration

$$[(za)_{2k-1}+(za)_{2k}]^{in} = u^6_{2k-2}n^6_{2k-2}f_k(ZA)^+_{in}$$

$$(ZA)^+_{in} = -3(F_2B_1-B_2\bar{E}_1)/2H$$

Fifth-Order Coma

$$[(zb)_{2k-1}+(zb)_{2k}]^{in} = u^5_{2k-2}n^5_{2k-2}\bar{y}^{\#}_{2k-2}(ZB)^+_{in}$$

$$(ZB)^+_{in} = -[B_1(P_2H^2+4C_2)+F_2(5F_1-4\bar{E}_1)-B_2(2P_1H^2+5\bar{C}_1)]/2H$$

$$[(zc)_{2k-1}+(zc)_{2k}]^{in} = u^5_{2k-2}n^5_{2k-2}\bar{y}^{\#}_{2k-2}(ZC)^+_{in}$$

$$(ZC)^+_{in} = -[B_1(P_2H^2-2C_2)+2F_2(2F_1-\bar{E}_1)-B_2(P_1H^2+4\bar{C}_1)]/2H$$

Table 7.8. (cont.)

Oblique Spherical Aberration

$$[(zd)_{2k-1}+(zd)_{2k}]^{in} = (u^4_{2k-2}n^4_{2k-2}\bar{y}^{\#2}_{2k-2}/f_k)(ZD)^+_{in}$$

$$(ZD)^+_{in} = -[E_2B_1+C_2(4F_1-\bar{E}_1)+F_2(C_1-4\bar{C}_1-2P_1H^2)]/2H$$

$$[(ze)_{2k-1}+(ze)_{2k}]^{in} = (u^4_{2k-2}n^4_{2k-2}\bar{y}^{\#2}_{2k-2}/f_k)(ZE)^+_{in}$$

$$(ZE)^+_{in} = -[(P_2H^2+C_2)(2F_1-\bar{E}_1)+F_2(P_1H^2+3C_1-2\bar{C}_1)-2B_2\bar{F}_1]/2H$$

$$[(zf)_{2k-1}+(zf)_{2k}]^{in} = (u^4_{2k-2}n^4_{2k-2}\bar{y}^{\#2}_{2k-2}/f_k)(ZF)^+_{in}$$

$$(ZF)^+_{in} = -[F_1(2C_2+P_2H^2)+F_2(C_1-2\bar{C}_1)-B_2\bar{F}_1]/2H$$

Fifth-Order "Petzval"

$$[(zn)_{2k-1}+(zn)_{2k}]^{in} = (u^2_{2k-2}n^2_{2k-2}\bar{y}^{\#4}_{2k-2}/f^3_k)(ZN)^+_{in}$$

$$(ZN)^+_{in} = -[E_2(P_1H^2-2C_1)+P_2(4E_1-\bar{F}_1)+2C_2(E_1+\bar{F}_1)-2F_2\bar{B}_1]/2H$$

Elliptical Coma

$$[(zg)_{2k-1}+(zg)_{2k}]^{in} = (u^3_{2k-2}n^3_{2k-2}\bar{y}^{\#3}_{2k-2}/f^2_k)(ZG)^+_{in}$$

$$(ZG)^+_{in} = -[3E_2F_1-(P_2H^2+C_2)(P_1H^2+\bar{C}_1)+2C_2(C_1+\bar{C}_1)+F_2(P_1H^2-2\bar{F}_1)-B_2\bar{B}_1]/2H$$

$$[(zh)_{2k-1}+(zh)_{2k}]^{in} = (u^3_{2k-2}n^3_{2k-2}\bar{y}^{\#3}_{2k-2}/f^2_k)(ZH)^+_{in}$$

$$(ZH)^+_{in} = -[E_2F_1+(P_1H^2+3C_2)(3C_1-\bar{C}_1+P_1H^2)-C_2(P_1H^2+C_1)+F_2(E_1-8\bar{F}_1)-B_2\bar{B}_1]/2H$$

$$[(zk)_{2k-1}+(zk)_{2k}]^{in} = (u^3_{2k-2}n^3_{2k-2}\bar{y}^{\#3}_{2k-2}/f^2_k(ZK)^+_{in}$$

$$(ZK)^+_{in} = -[E_2F_1+(P_2H^2+C_2)(3C_1+P_1-\bar{C}_1)+C_2(C_1+P_1H^2)-B_2\bar{B}_1+F_2(E_1+4\bar{F}_1)]/2H$$

Fifth-Order Image Distortion

$$[(zp)_{2k-1}+(zp)_{2k}]^{in} = (u_{2k-2}n_{2k-2}\bar{y}^{\#5}_{2k-2}/f^4_k)(ZP)^+_{in}$$

$$(ZP)^+_{in} = -[3E_1E_2-\bar{B}_1(P_2H^2+3C_2)]/2H$$

Fifth-Order Pupil Distortion

$$[(zq)_{2k-1}+(zq)_{2k}]^{in} = u^5_{2k-2}n^5_{2k-2}\bar{y}^{\#}_{2k-2}(ZQ)^+_{in}$$

$$(ZQ)^+_{in} = -[B_1(P_2H^2+3\bar{C}_2)-\bar{E}_1\bar{E}_2]$$

Fifth-Order Astigmatism

$$[(zm)_{2k-1}+(zm)_{2k}]^{in} = (u^2_{2k-2}n^2_{2k-2}\bar{y}^{\#4}_{2k-2}/f^3_k)(ZM)^+_{in}$$

$$(ZM)^+_{in} = -[4E_2C_1+E_2P_1H^2-P_2F_1H^2+2C_2(E_1+2\bar{F}_1)-F_2\bar{B}_1]/2H$$

Table 7.9. Extrinsic external contributions

Fifth-Order Spherical Aberration

$$[(za)_{2k-1} + (za)_{2k}]^{ex} = 0$$

Fifth-Order Coma

$$[(zb)_{2k-1} + (zb)_{2k}]^{ex} = u_{2k-2}^2 n_{2k-2}^2 F_k \sum^{k-1} u_{2m}^3 n_{2m}^3 \bar{y}_{2m}^{\#}(5F_m - 4\bar{E}_m)/2H$$

$$[(zc)_{2k-1} + (zc)_{2k}]^{ex} = 2u_{2k-2}^2 n_{2k-2}^2 F_k \sum^{k-1} u_{2m}^3 n_{2m}^3 \bar{y}_{2m}^{\#}(2F_m - \bar{E}_m)/2H$$

Oblique Spherical Aberration

$$[(zd)_{2k-1} + (zd)_{2k}]^{ex} = 2u_{2k-2}^2 n_{2k-2}^2 F_k \sum^{k-1} (u_{2m}^2 n_{2m}^2 \bar{y}_{2m}^{\#}/f_m)(P_m H^2 - 2\bar{C}_m)/2H$$

$$[(ze)_{2k-1} + (ze)_{2k}]^{ex} = (u_{2k-2}^2 n_{2k-2}^2 \bar{y}_{2k-2}^{\#}/f_k)P_k H^2 \sum^{k-1} u_{2m}^3 n_{2m}^3 \bar{y}_{2m}^{\#}(2F_m - E_m)/2H + (u_{2k-2}^2 n_{2k-2}^2 F_k) \sum^{k-1} (u_{2m}^2 n_{2m}^2 \bar{y}_{2m}^{\#2}/f_m)(P_m - 2\bar{C}_m)/2H$$

$$[(zf)_{2k-1} + (zf)_{2k}]^{ex} = (u_{2k-2}^2 n_{2k-2}^2 \bar{y}_{2k-2}^{\#}/f_k)P_k \sum^{k-1} u_{2m} n_{2m} \bar{y}_{2m}^{\#} F_m/2H - 2u_{2k-2}^2 n_{2k-2}^2 F_k \sum^{k-1} (u_{2m}^2 n_{2m}^2 \bar{y}_{2m}^{\#}/f_m)\bar{C}_m/2H$$

Elliptical Coma

$$[(zg)_{2k-1} + (zg)_{2k}]^{ex} = 3(\bar{y}_{2k-2}^{\#2}/f_k^2) \sum^{k-1} u_{2m}^3 n_{2m}^3 \bar{y}_{2m}^{\#} F_m/2H - (u_{2k-2}^2 n_{2k-2}^2 \bar{y}_{2k-2}^{\#}/f_k)P_k H \sum^{k-1} (u_{2m} n_{2m} \bar{y}_{2m}^{\#}/f_m)(P_m H^2 + \bar{C}_m)/2$$
$$+ u_{2k-2}^2 n_{2k-2}^2 F_k \sum^{k-1} (u_{2m} n_{2m} \bar{y}_{2m}^{\#3}/f_m^2)(E_m - 2\bar{F}_m)/2H$$

$$[(zh)_{2k-1} + (zh)_{2k}]^{ex} = 3(\bar{y}_{2k-2}^{\#2}/f_k^2)E_k \sum^{k-1} u_{2m}^3 n_{2m}^3 \bar{y}_{2m}^{\#} F_k/2H + (u_{2k-2}^2 n_{2k-2}^2 \bar{y}_{2k-2}^{\#}/f_k)P_k \sum^{k-1} (u_{2m}^2 n_{2m}^2 \bar{y}_{2m}^{\#2}/f_m)(P_m H^2 - \bar{C}_m)/2H$$
$$+ u_{2k-2}^2 n_{2k-2}^2 F_k \sum^{k-1} (u_{2m} n_{2m} \bar{y}_{2m}^{\#3}/f_m^2)(E_m - 8\bar{F}_m)/2H$$

Table 7.9. (cont.)

$$[(zk)_{2k-1} + (zk)_{2k}]^{ex} = (\bar{y}^{\#2}_{2k-2}/f^2_k)E_k \sum^{k-1} u^3_{2m}n^3_{2m}\bar{y}^{\#}_{2m}F_m/2H + (u^3_{2k-2}n^3_{2k-2}\bar{y}^{\#}_{2k-2}/f_k)P_kH^2 \sum^{k-1} (u^2_{2m}n^2_{2m}\bar{y}^{\#2}_{2m}/f_m)\bar{C}_m/2H$$

$$+ u^2_{2k-2}n^2_{2k-2}F_k \sum^{k-1} (u_{2m}n_{2m}\bar{y}^{\#}_{2m}/f^2_m)(E_m - 4\bar{F}_m)/2H$$

Fifth-Order Astigmatism

$$[(zm)_{2k-1} + (zm)_{2k}]^{ex} = (u^2_{2k-2}n^2_{2k-2}\bar{y}^{\#4}_{2k-2}/f^2_k)E_kH^2 \sum^{k-1} (1/f_m)P_m/2H - (u_{2k-2}n_{2k-2}\bar{y}^{\#}_{2k-2}/f_k)P_kH^2 \sum^{k-1} (\bar{y}^{\#4}_{2m}/f^3_m)\bar{B}_m/2H$$

$$- 2u^2_{2k-2}n^2_{2k-2}F_k \sum^{k-1} (\bar{y}^{\#4}_{2m}/f^3_m)\bar{B}_m/2H$$

Fifth-Order "Petzval"

$$[(zn)_{2k-1} + (zn)_{2k}]^{ex} = (u^2_{2k-2}n^2_{2k-2}\bar{y}^{\#4}_{2k-2}/f^2_k)E_kH^2 \sum^{k-1} (1/f_m)P_m/2H - 2u^2_{2k-2}n^2_{2k-2}F_k \sum^{k-1} (\bar{y}^{\#4}_{2m}/f^3_m)\bar{B}_m/2H$$

$$- (u_{2k-2}n_{2k-2}\bar{y}^{\#}_{2k-2}/f_k)P_kH^2 \sum^{k-1} (u_{2m}n_{2m}\bar{y}^{\#3}_{2m}/f^2_m)(4E_m - \bar{F}_m)/2H$$

Fifth-Order Image Distortion

$$[(zp)_{2k-1} + (zp)_{2k}]^{ex} = 3(\bar{y}^{\#2}_{2k-2}/f_k)E_k \sum^{k-1} (u_{2m}n_{2m}\bar{y}^{\#}_{2m}/f^2_m)E_m/2H - (u_{2k-2}n_{2k-2}\bar{y}^{\#}_{2k-2}/f_k) \sum^{k-1} (\bar{y}^{\#4}_{2m}/f^3_m)\bar{B}_k/2H$$

Fifth-Order Pupil Distortion

$$[(zq)_{2k-1} + (zq)_{2k}]^{ex} = -u^2_{2k-2}n^2_{2k-2}\bar{E}_k \sum^{k-1} u^3_{2m}n^3_{2m}\bar{y}^{\#}_{2m}\bar{E}_m/2H$$

7.4 The Backward Orientation

In many respects, this section is a mirror image of the last. Again, we use the intrinsic contributions from Tables 7.3,4 and the extrinsic contributions in Table 7.5 to obtain the canonical forms of the fifth-order aberrations for a backward module. Again we consider the kth module in an optical system. Its first surface will have the index 2k-1; its second, 2k.

But we will spoil the symmetry. Instead of calculating fifth-order spherical aberration, we will do fifth-order astigmatism in order to have a closer look at the extrinsic external contributions. We use Table 7.4, to find

$$(ym)_j = \bar{a}_j (se)_j / 2n_{j-1} \quad , \tag{7.4.1}$$

and Table 7.3 for

$$(se)_j = n_{j-1}(\mu_j - 1)\{(xh)_j(xb)_j + (xf)_j(xk)_j + y_j(a_j/n_j - 1)[\bar{a}_j(xc_j)/n_{j-1}$$
$$-\bar{u}_j(xg)_j]\}/4 \quad . \tag{7.4.2}$$

Also from Table 7.3, we get

$$(xh)_j = [\bar{y}_j^2 c_j a_j + y_j \bar{a}_j(\bar{u}_j + \bar{u}_{j-1})]/n_{j-1}$$

$$(xb)_j = 3a_j \bar{a}_j / n_j n_{j-1} + 2u_j \bar{u}_j - 3u_{j-1}\bar{u}_{j-1}$$

$$(xf)_j = [\bar{a}_j(2u_{j-1} - u_j) + a_j \bar{u}_{j-1}]/n_{j-1}$$

$$(xk)_j = \bar{y}_j^2 c_j u_{n-1} - y_j \bar{a}_j(\bar{u}_j + \bar{u}_{j-1})/n_j$$

$$(xc)_j = 3\bar{a}_j^2 / n_j n_{j-1} + 2\bar{u}_j^2 - 3\bar{u}_{j-1}^2$$

$$(xg)_j = \bar{a}_j(3\bar{u}_{j-1} - \bar{u}_j)/n_{j-1} \quad . \tag{7.4.3}$$

We operate first on the expressions in (7.4.3), using the canonical variables derived for the backward module that are found in Tables 6.5,6 and (6.6.2). The results of these calculations, along with others, are found in Table 7.10. These results are then applied to (7.4.2), which yields the canonical version of $(se)_{2k-1}$, which is also found in Table 7.10. Finally, we move to (7.4.1) and obtain, for the intrinsic contribution for fifth-order astigmatism at the first surface of the kth module

$$(ym)_{2k-1} = (y''^{\#2}_{2k-1} n^4_{2k-2} \bar{u}^{-4}_{2k-2}/f)(YM)^-_2 \quad , \tag{7.4.4}$$

where

$$(YM)^-_2 = -\bar{A}_2(SE)^-_2/N_2 \quad . \tag{7.4.5}$$

Table 7.10. Canonical intrinsic fifth-order aberrations, backward module. First surface only

W_{2k-1}	$= (y''^{\#2}_{2k-1}/f^2_k)W^-_2$	W^-_2	$= [(A_2/N_1)^2(1+\mu^2_2)+U^2_1-3U^2_2]/8$
$(xa)_{2k-1}$	$= (y^{\#}_{2k-1}/f^2_k)(XA)^-_2$	$(XA)^-_2$	$= 3A^2_2/N_1N_2 + 2U^2_1 - 3U^2_2$
$(xb)_{2k-1}$	$= (y^{\#}_{2k-1}n_{2k-2}\bar{u}_{2k-2}/f^2_k)(XB)^-_2$	$(XB)^-_2$	$= A_2\bar{A}_2/N_1N_2 + 2U_1\bar{U}_1 - 3U_2\bar{U}_2$
$(xc)_{2k-1}$	$= (y''^{\#2}_{2k-1}/f^2_k)(XC)^-_2$	$(XC)^-_2$	$= 3\bar{A}^2_2/N_1N_2 + 2\bar{U}^2_1 - 3\bar{U}^2_2$
$(xe)_{2k-1}$	$= (y''^{\#2}_{2k-1}/f^2_k)(XE)^-_2$	$(XE)^-_2$	$= A_2(3U_2-U_1)/N_2$
$(xf)_{2k-1}$	$= (y^{\#}_{2k-1}n_{2k-2}\bar{u}_{2k-2}/f_k)(XF)^-_2$	$(XF)^-_2$	$= [\bar{A}_2(2U_2-U_1)+A_2\bar{U}_2]/N_2$
$(xg)_{2k-1}$	$= n^2_{2k-2}\bar{u}^2_{2k-2}(XG)^-_2$	$(XG)^-_2$	$= \bar{A}_2(3\bar{U}_2-\bar{U}_1)/N_2$
$(xh)_{2k-1}$	$= (y''^{\#2}_{2k-1}n_{2k-2}\bar{u}_{2k-2}/f_k)(XH)^-_2$	$(XH)^-_2$	$= [\bar{Y}^2_2C_2A_2+Y_2\bar{A}_2(\bar{U}_1+\bar{U}_2)]/N_2$
$(xk)_{2k-1}$	$= (y''^{\#2}_{2k-1}n_{2k-2}\bar{u}_{2k-2}/f_k)(XK)^-_2$	$(XK)^-_2$	$= \bar{Y}^2_2C_2U_2 - Y_2\bar{A}_2(\bar{U}_1+\bar{U}_2)/N_1$
$(xm)_{2k-1}$	$= (y''^{\#2}_{2k-1}n_{2k-2}\bar{u}_{2k-2}/f_k)(XM)^-_2$	$(XM)^-_2$	$= [Y^2_2C_2\bar{A}_2-\bar{Y}_2A_2(U_1+U_2)]/N_2$
$(xn)_{2k-1}$	$= (y''^{\#2}_{2k-1}n_{2k-2}\bar{u}_{2k-2}/f_k)(XN)^-_2$	$(XN)^-_1$	$= Y^2_2C_2\bar{U}_2 - \bar{Y}_2A_2(U_1+U_2)/N_1$
$(sa)_{2k-1}$	$= (y^{\#5}_{2k-1}/f^4_k)(SA)^-_2$	$(SA)^-_2$	$= 3W^-_2S_2Y_2\mu^2_2A_2/2$
$(sb)_{2k-1}$	$= (y^{\#4}_{2k-1}n_{2k-2}\bar{u}_{2k-2}/f^3_k)(SB)^-_2$	$(SB)^-_2$	$= Y_2S_2N^2_1\bar{A}_2(XA)^-_2 + A_2(XB)^-_2/N_1$
			$\quad - \bar{U}_1(XE)^-_2 + U_1(XF)^-_2/4$
$(sc)_{2k-1}$	$= (y^{\#3}_{2k-1}n^2_{2k-2}\bar{u}^2_{2k-2}/f^2_k)(SC)^-_2$	$(SC)^-_2$	$= (N_2/8)(\mu_1-1)[(XH)^-_2(XA)^-_2+(XE)^-_2+(XK)^-_2$
			$\quad +Y_2(A_2/N_2+U_1)A_2(XC)^-_2/N_2-U_1(XG)^-_2]$
$(sd)_{2k-1}$	$= (y^{\#3}_{2k-1}n^2_{2k-2}\bar{u}^2_{2k-2}/f^2_k)(SD)^-_2$	$(SD)^-_2$	$= Y_2S_2N^2_1\bar{A}_2(XB)^-_2/N_2 - \bar{U}_1(XF)^-_2/2$
$(se)_{2k-1}$	$= (y''^{\#2}_{2k-1}n^3_{2k-2}\bar{u}^3_{2k-2}/f_k)(SE)^-_2$	$(SE)^-_2$	$= (N_2/8)(\mu_1-1)[(XH)^-_2(XB)^-_2+(XF)^-_2(XK)^-_2$
			$\quad +Y_2(\bar{A}_2/N_2+U_1)\bar{A}_2(XC)^-_2/N_2-\bar{U}_2(XG)^-_2]$
$(sf)_{2k-1}$	$= y^{\#}_{2k-1}n^4_{2k-2}\bar{u}_{2k-2}(SF)^-_2$	$(SF)^-_2$	$= (N_2/8)(\mu_2-1)[(XH)^-_2(XC)^-_2+(XG)^-_2(XK)^-_2]$
$(sg)_{2k-1}$	$= (y''^{\#2}_{2k-1}n_{2k-2}\bar{u}_{2k-2}/f_k)(SG)^-_2$	$(SG)^-_2$	$= (N_2/8)(\mu_2-1)[(XM)^-_2(XA)^-_2+(XE)^-_2(XN)^-_2]$

The expression $(YM)_2^-$ in (7.4.5) is now the canonical form of the contribution to fifth-order astigmatism from the first surface of the kth module.
The subscript is *2* rather than *1* because this is the second surface of a module in its normal or forward orientation. The superior (-) denotes, further,
that the calculation is for a backward module. The results of this calculation
and those for the other fifth-order aberrations appear in Table 7.11. Here
we show the quantities for the first surface only. Those for the second surface follow in an obvious way, as in the case of astigmatism

$$(ym)_{2k} = (y_{2k-1}^{\#2} n_{2k-2}^4 \bar{u}_{2k-2}^{-4}/f_k)(YM)_1^- , \tag{7.4.6}$$

where

$$(YM)_1^- = -\bar{A}_1 (SE)_1^-/N_1 . \tag{7.4.7}$$

Now we proceed to the extrinsic contributions. We begin with the expression for fifth-order astigmatism from Table 7.5

$$(zm)_j = \left[e_j \sum^{j-1} (4c_k+p_k h^2)-p_j h^2 \sum^{j-1} \bar{b}_k+2c_j \sum^{j-1} (e_k-2\bar{b}_k)-2b_j \sum^{j-1} \bar{b}_k \right]/4h .$$

The contributions from each surface of the module are added together to yield

$$(zm)_{2k-1} + (zm)_{2k} = \left[e_{2k-1} \sum^{2k-1} (4c_m+P_m h^2)-P_{2k-1}h^2 \sum^{2k-1} \bar{b}_m \right.$$
$$+ 2c_{2k-1} \sum^{2k-1} (e_m-2\bar{b}_m)-2b_{2k-1} \sum^{2k-1} \bar{b}_m$$
$$+ e_{2k} \sum^{2k} (4c_m+P_m h^2)-P_{2k}h^2 \sum^{2k} \bar{b}_m$$
$$\left. + 2c_{2k} \sum^{2k} (e_m-2\bar{b}_m)-2b_{2k} \sum^{2k} \bar{b}_m \right] / 4h$$
$$= \left[4e_{2k}c_{2k-1}+e_{2k}P_{2k-1}h^2-P_{2k}h^2\bar{b}_{2k-1} \right.$$
$$+ 2c_{2k}(e_{2k-1}-2\bar{b}_{2k-1})-2b_{2k}\bar{b}_{2k-1}$$
$$+ (e_{2k}+e_{2k-1}) \sum^{2k-1} P_m h^2-(P_{2k}+P_{2k-1})h^2 \sum^{2k-1} \bar{b}_m$$
$$\left. + (c_{2k}+c_{2k-1}) \sum^{2k-1} (e_k-2\bar{b}_k)-2(b_{2k}+b_{2k-1}) \sum^{2k-1} \bar{b}_k \right] / 4h . \tag{7.4.8}$$

Table 7.11. Canonical intrinsic fifth-order aberration, backward module.

Fifth-Order Spherical Aberration

$(ya)_{2k-1} = (y_{2k-1}^{\#6}/f_k^5)(YA)_2^-$ $(YA)_2^- = -2A_2(SA)_2^-/N_2$

Fifth-Order Coma

$(yb)_{2k-1} = (y_{2k-1}^{\#5}n_{2k-2}\bar{u}_{2k-2}/f_k^4)(YB)_2^-$ $(YB)_2^- = -2\bar{A}_2(SA)_2^-/N_2 + (YC)_2^-$

$(yc)_{2k-1} = (y_{2k-1}^{\#5}n_{2k-2}\bar{u}_{2k-2}/f_k^4)(YC)_2^-$ $(YC)_2^- = -A_2(SB)_2^-/N_2$

Oblique Spherical Aberration

$(yd)_{2k-1} = (y_{2k-1}^{\#4}n_{2k-2}^2\bar{u}_{2k-2}^2/f_k^3)(YD)_2^-$ $(YD)_2^- = -2\bar{A}_2(SB)_2^-/N_2$

$(ye)_{2k-1} = (y_{2k-1}^{\#4}n_{2k-2}^2\bar{u}_{2k-2}^2/f_k^3)(YE)_2^-$ $(YE)_2^- = -2A_2(SC)_2^-/N_2$

$(yf)_{2k-1} = (y_{2k-1}^{\#4}n_{2k-2}\bar{u}_{2k-2}/f_k^3)(YF)_2^-$ $(YF)_2^- = -2A_2(SD)_2^-/N_2$

Elliptical Coma

$(yg)_{2k-1} = (y_{2k-1}^{\#3}n_{2k-2}^3\bar{u}_{2k-2}^3/f_k^2)(YG)_2^-$ $(YG)_2^- = -2\bar{A}_2(SC)_2^-/N_2$

$(yh)_{2k-1} = (y_{2k-1}^{\#3}n_{2k-2}^3\bar{u}_{2k-2}^3/f_k^2)(YH)_2^-$ $(YH)_2^- = -2\bar{A}_2(SD)_2^-/N_2 + (YK)_2^-$

$(yk)_{2k-1} = (y_{2k-1}^{\#3}n_{2k-2}^3\bar{u}_{2k-2}^3/f_k^2)(YK)_2^-$ $(YK)_2^- = -A_2(SE)_2^-/N_2$

Fifth-Order Astigmatism

$(ym)_{2k-1} = (y_{2k-1}^{\#2}n_{2k-2}^4\bar{u}_{2k-2}^4/f_k)(YM)_2^-$ $(YM)_2^- = -\bar{A}_2(SE)_2^-/N_2$

Fifth-Order "Petzval"

$(yn)_{2k-1} = (y_{2k-1}^{\#2}n_{2k-2}^4\bar{u}_{2k-2}^4/f_k)(YN)_2^-$ $(YN)_2^- = -2A_2(SF)_2^-/N_2 + (YM)_2^-$

Fifth-Order Image Distortion

$(yp)_{2k-1} = (y_{2k-1}^{\#}n_{2k-2}^5\bar{u}_{2k-2}^5)(YP)_2^-$ $(YP)_2^- = -2\bar{A}_2(SF)_2^-/N_2$

Fifth-Order Pupil Distortion

$(yq)_{2k-1} = (y_{2k-1}^{\#4}n_{2k-2}\bar{u}_{2k-2}/f_k^2)(YQ)_2^-$ $(YQ)_2^- = -2A_2(SG)_2^-/N_2$

Once again we separate the extrinsic internal and extrinsic external contributions. The internal, first, is

$$[(zm)_{2k-1} + (zm)_{2k}]^{in} = \Big[4e_{2k}c_{2k-1} + e_{2k}P_{2k-1}h^2 - P_{2k}h^2\bar{\delta}_{2k-1}$$
$$+ 2c_{2k}(e_{2k-1} - 2\bar{\delta}_{2k-1}) - 2\delta_{2k}\bar{b}_{2k-1}\Big]/4h$$
$$= (y_{2k-1}^{\#2}n_{2k-2}^4\bar{u}_{2k-2}^4/f_k)(ZM)_{in}^- \quad , \tag{7.4.9}$$

where

$$(ZM)_{in}^{-} = -\left[E_1(P_2H^2-2C_2)+P_1H^2(4E_2-\bar{F}_2)+2C_1(E_2+\bar{F}_1)-2F_1\bar{B}_2\right]/4H \quad . \qquad (7.4.10)$$

These two results appear in Table 7.12, as do the results of calculating the other extrinsic internal contributions.

We finally turn to the calculation of the extrinsic external contributions. Returning to (7.4.8), we pick up the summed terms found there to obtain

$$[(zm)_{2k-1}+(zm)_{2k}]^{ex} = \left[(e_{2k}+e_{2k-1})\sum_{m}^{2k-1}P_mh^2-(P_{2k}+P_{2k-1})h^2\sum^{2k-1}\bar{\delta}_m\right.$$
$$\left.+(c_{2k}+c_{2k-1})\sum^{2k-1}(e_k-2\bar{\delta}_k)-2(\delta_{2k}+\delta_{2k-1})\sum^{2k-1}\bar{\sigma}_k\right]/4h \quad .$$
$$(7.4.11)$$

Note first of all that $c_{2k}+c_{2k-1}=0$, because third-order astigmatism vanishes over a module and one term disappears.

We proceed, as in the previous section, to substitute the canonical versions of the appropriate third-order aberrations and obtain

$$[(zm)_{2k-1}+(zm)_{2k}]^{ex} = -n_{2k-2}^2\bar{u}_{2k-2}^2E_k\sum^{k-1}(y_{2m+1}^{\#}n_{2m}\bar{u}_{2m}/f_m)P_m/4$$
$$+ (y_{2k-1}^{\#}n_{2k-2}\bar{u}_{2k-2}/f_k)H\sum^{k-1}y_{2m+1}^{\#}n_{2m}\bar{u}_{2m}\bar{F}_m/4$$
$$+ (y_{2k-2}^{\#2}/f_k^2)F_k\sum^{k-1}n_{2m}\bar{u}_{2m}f_m\bar{B}_k/4H \quad , \qquad (7.4.12)$$

where E_k, P_m, etc. represent the canonical third-order contributions over the entire k^{th} and m^{th} module, as defined in (6.3.13-19).

The results of this and similar calculations are collected in Table 7.13.

Table 7.12. Extrinsic internal contributions. Fifth-order aberrations

---

Fifth-Order Spherical Aberration

$[(za)_{2k-1}+(za)_{2k}]^{in} = (y_{2k-1}^{\#6}/f_k^5)(ZA)_{in}^{-}$

$(ZA)_{in}^{-} = -3(F_1B_2-B_1\bar{E}_2)/2H$

Fifth-Order Coma

$[(zb)_{2k-1}+(zb)_{2k}]^{in} = (y_{2k-1}^{\#5}n_{2k-2}\bar{u}_{2k-2}/f_k^4)(ZB)_{in}^{-}$

$(ZB)_{in}^{-} = -[B_2(P_1H^2+4C_1)+F_1(5F_2-4\bar{E}_1)-B_1(2P_2H^2+5\bar{C}_2)]/2H$

$[(zc)_{2k-1}+(zc)_{2k}]^{in} = (y_{2k-1}^{\#5}n_{2k-2}\bar{u}_{2k-2}/f_k^4)(ZC)_{in}^{-}$

$(ZC)_{in}^{-} = -[B_2(P_1H^2-2C_1)+2F_1(2F_2-4\bar{E}_1)-B_1(P_2H^2+4\bar{C}_2)]/2H$

Table 7.12. (cont.)

Oblique Spherical Aberration

$$[(zd)_{2k-1}+(zd)_{2k}]^{in} = (y_{2k-1}^{\#4} n_{2k-2}^2 \bar{u}_{2k-2}/f_k^3)(ZD)_{in}^-$$

$$(ZD)_{in}^- = -[E_1 B_2+(P_1 H^2+C_1)(2F_2-\bar{E}_2)+F_2(P_2 H^2+3C_2-2\bar{C}_2)-3B_1\bar{F}_2]/2H$$

$$[(ze)_{2k-1}+(zd)_{2k}]^{in} = (y_{2k-1}^{\#4} n_{2k-2}^2 \bar{u}_{2k-2}^2/f_k^3)(ZE)_{in}^-$$

$$(ZE)_{in}^- = -[E_1 B_2+(P_1 H^2+C_1)(2F_2-\bar{E}_2)+F_2(P_2 H^2+3C_2-2\bar{C}_2)-3B_1\bar{F}_2]/2H$$

$$[(zf)_{2k-1}+(zf)_{2k}]^{in} = (y_{2k-1}^{\#4} n_{2k-2}^2 \bar{u}_{2k-2}^2/f_k^3)(ZF)_{in}^-$$

$$(ZF)_{in}^- = -2[F_2(C_1+P_1 H^2)+F_1(C_2-2\bar{C}_2)+B_1\bar{F}_2]/2H$$

Fifth-Order Astigmatism

$$[(zm)_{2k-1}+(zm)_{2k}]^{in} = (y_{2k-1}^{\#2} n_{2k-2}^4 \bar{u}_{2k-2}^{-4}/f_k)(ZM)_{in}^-$$

$$(ZM)_{in}^- = -[E_1(4C_2+P_2 H^2)-P_1 H^2\bar{F}_2+2C_1(E_2+\bar{F}_1)-2F_1\bar{B}_2]/4H$$

Fifth-Order "Petzval"

$$[(zn)_{2k-1}+(zn)_{2k}]^{in} = (y_{2k-1}^{\#2} n_{2k-2}^4 \bar{u}_{2k-2}^{-4}/f_k)(ZN)_{in}^-$$

$$(ZN)_{in}^- = -[E_1(P_2 H^2-2C_2)+P_1 H^2(4E_2-\bar{F}_2)+2C_1(E_2+\bar{F}_1)-2F_1\bar{B}_2]/4H$$

Elliptical Coma

$$[(zg)_{2k-1}+(zg)_{2k}]^{in} = (y_{2k-1}^{\#3} n_{2k-2}^3 \bar{u}_{2k-2}^{-3}/f_k^2)(ZG)_{in}^-$$

$$(ZG)_{in}^- = -[3E_1 F_2-(P_1 H^2+C_1)(P_2 H^2+\bar{C}_2)+2C_1(C_2-\bar{C}_2)+F_1(E_2-2\bar{F}_2)-\bar{B}_2 B_1]/2H$$

$$[(zh)_{2k-1}+(zh)_{2k}]^{in} = (y_{2k-1}^{\#3} n_{2k-2}^3 \bar{u}_{2k-2}^{-3}/f_k^2)(ZH)_{in}^-$$

$$(ZH)_{in}^- = -[E_1 F_2+(P_1 H^2+3C_1)(3C_2-\bar{C}_2+P_2 H^2)-C_1(P_2 H^2+C_2)+F_1(E_2-8\bar{F}_2)-B_1\bar{B}_2]/H$$

$$[(zk)_{2k-1}+(zk)_{2k}]^{in} = (y_{2k-1}^{\#3} n_{2k-2}^3 \bar{u}_{2k-2}^{-3}/f_k^2)(ZK)_{in}^-$$

$$(ZK)_{in}^- = -[E_1 E_2+(P_1 H^2+C_1)(3C_2+P_2 H^2-\bar{C}_2)+C_1(C_2+P_2 H^2)+F_1(E_2-4\bar{F}_2)-B_1\bar{B}_2]/4H$$

Fifth-Order Distortion

$$[(zp)_{2k-1}+(zp)_{2k}]^{in} = y_{2k-1}^{\#} n_{2k-2}^5 \bar{u}_{2k-2}^{-5}(ZP)_{in}^-$$

$$(ZP)_{in}^- = -[3E_1 E_2+\bar{B}_2(P_1 H^2+3C_1)]/2H$$

Fifth-Order Pupil Distortion

$$[(zq)_{2k-1}+(zq)_{2k}]^{in} = (y_{2k-1}^{\#5} n_{2k-2} \bar{u}_{2k-2}/f_k^4)(ZQ)_{in}^-$$

$$(ZQ)_{in}^- = [B_2(P_1 H^2+3\bar{C}_1)-\bar{E}_2\bar{E}_1]/2H$$

Table 7.13. Extrinsic external contribution. Fifth-order aberrations

---

Fifth-Order Spherical Aberration

$[(za)_{2k+1} + (za)_{2k}]^{ex} = 0$

Fifth-Order Coma

$[(zb)_{2k-1} + (zb)_{2k}]^{ex} = -(y_{2k-1}^{\#2}/f_k^2)F_k \sum\limits^{k-1} (y_{2m+1}^{\#3} n_{2m} \bar{u}_{2m}/f_m^2)(5F_m - 4\bar{E}_m)/2H$

$[(zc)_{2k-1} + (zc)_{2k}]^{ex} = -(y_{2k-1}^{\#2}/f_k^2)F_k \sum\limits^{k-1} (y_{2m+1}^{\#3} n_{2m} \bar{u}_{2m}/f_m^2)(2F_m - \bar{E}_m)/2H$

Oblique Spherical Aberration

$[(zd)_{2k-1} + (zd)_{2k}]^{ex} = (y_{2k-1}^{\#2} n_{2k-2} \bar{u}_{2k-2}/f_k^2)F_k \sum\limits^{k-1} (y_{2m+1}^{\#2} n_{2m} \bar{u}_{2m}/f_m)(2\bar{C}_m - P_m H^2)/H$

$[(ze)_{2k-1} + (ze)_{2k}]^{ex} = (y_{2k-1}^{\#} n_{2k-2} \bar{u}_{2k-2}/f_k)P_k H \sum\limits^{k-1} (y_{2m+1}^{\#3} n_{2m} \bar{u}_{2m}/f_m^2)(2F_m - \bar{E}_m)/2$

$\qquad\qquad +(y_{2k-1}^{\#2}/f_k^2)F_k \sum\limits^{k-1} (y_{2m+1}^{\#2} n_{2m}^2 \bar{u}_{2m}^2/f_m)(P_m H^2 - 2\bar{C}_m)/2H$

$[(zf)_{2k-1} + (zf)_{2k}]^{ex} = -2(y_{2k-1}^{\#2}/f_k^2)F_k \sum\limits^{k-1} (y_{2m+1}^{\#2} n_{2m}^2 \bar{u}_{2m}^2/f_m)\bar{C}_m/H$

$\qquad\qquad -2(y_{2k-1}^{\#} n_{2k-2} \bar{u}_{2k-2}/f_k)P_k H \sum\limits^{k-1} (y_{2m+1}^{\#3} n_{2m} \bar{u}_{2m}/f_m^2)F_m$

Elliptical Coma

$[(zg)_{2k-1} + (zg)_{2k}]^{ex} = -3n_{2k-2}^2 \bar{u}_{2k-2}^2 E_k \sum\limits^{k-1} (y_{2m+1}^{\#3} n_{2m} \bar{u}_{2m}/f_m^2)F_m/2H$

$\qquad\qquad +(y_{2k-1}^{\#} n_{2k-2} \bar{u}_{2k-2}/f_k)P_k \sum\limits^{k-1} (y_{2m+1}^{\#2} n_{2m}^2 \bar{u}_{2m}^2/f_m)(P_m H^2 + \bar{C}_m)/2$

$\qquad\qquad -(y_{2k-1}^{\#2}/f_k^2)F_k \sum\limits^{k-1} y_{2m+1}^{\#} n_{2m}^3 \bar{u}_{2m}^3 (E_m - 2\bar{F}_m)/2H$

$[(zh)_{2k-1} + (zh)_{2k}]^{ex} = -3n_{2k-2}^2 \bar{u}_{2k-2}^2 E_k \sum\limits^{k-1} (y_{2m+1}^{\#3} n_{2m} \bar{u}_{2m}/f_m^2)F_m/H$

$\qquad\qquad +(y_{2k-1}^{\#} n_{2k-2} \bar{u}_{2k-2}/f_k)P_k \sum\limits^{k-1} (y_{2m+1}^{\#2} n_{2m}^2 \bar{u}_{2m}^2/f_m)(\bar{C}_m - P_m H^2)$

$\qquad\qquad -(y_{2k-1}^{\#2}/f_k^2)F_k \sum\limits^{k-1} y_{2m+1}^{\#} n_{2m}^3 \bar{u}_{2m}^3 (E_m - 8\bar{F}_m)/H$

$[(zk)_{2k-1} + (zk)_{2k}]^{ex} = -n_{2k-2}^2 \bar{u}_{2k-2}^2 E_k \sum\limits^{k-1} (y_{2m+1}^{\#3} n_{2m} \bar{u}_{2m}/f_m^2)F_m/2H$

$\qquad\qquad -(y_{2k-1}^{\#} n_{2k-2} \bar{u}_{2k-2}/f_k)P_k H \sum\limits^{k-1} (y_{2m+1}^{\#2} n_{2m}^2 \bar{u}_{2m}^2/f_m)(P_m H^2 - \bar{C}_m)/2$

$\qquad\qquad -(y_{2k-1}^{\#2}/f_k^2)F_k \sum\limits^{k-1} y_{2m+1}^{\#} n_{2m}^3 \bar{u}_{2m}^3 (E_m - 4\bar{F}_m)/2H$

Table 7.13. (cont.)

Fifth-Order Astigmatism

$$[(zm)_{2k-1} + (zm)_{2k}]^{ex} = -n_{2k-2}^2 \bar{u}_{2k-2}^2 E_k H \sum^{k-1} (y_{2m+1}^{\#2} n_{2m}^2 \bar{u}_{2m}^2 / f_m) P_m / 4$$

$$+ (y_{2k-2}^{\#2} / f_k^2) F_k \sum^{k-1} n_{2m}^4 \bar{u}_{2m}^{-4} f_m \bar{B}_m / 2H$$

$$+ (y_{2k-1}^{\#} n_{2k-2} \bar{u}_{2k-2} / f_k) P_k H \sum^{k-1} y_{2m+1}^{\#} n_{2m}^3 \bar{u}_{2m}^{-3} \bar{F}_m / 4$$

Fifth-Order "Petzval"

$$[(zn)_{k-1} + (zn)_k]^{ex} = -n_{2k-2}^2 \bar{u}_{2k-2}^2 E_k H \sum^{k-1} (y_{2m+1}^{\#2} n_{2m}^2 \bar{u}_{2m}^2 / f_m) P_m / 4$$

$$- (y_{2k-1}^{\#} n_{2k-2} \bar{u}_{2k-2} / f_k) P_k H \sum^{k-1} y_{2m+1}^{\#} n_{2m}^3 \bar{u}_{2m}^{-3} (4E_m - \bar{F}_m) / 4$$

$$+ 2(y_{2k-1}^{\#2} / f_k^2) F_k \sum^{k-1} n_{2m}^4 \bar{u}_{2m}^{-4} f_m B_m / 4H$$

Fifth-Order Distortion

$$[(zp)_{k-1} + (zp)_k]^{ex} = -3n_{2k-2}^2 \bar{u}_{2k-2}^2 E_k \sum^{k-1} y_{2m+1}^{\#} n_{2m}^3 \bar{u}_{2m}^{-3} E_m / 2H$$

$$+ (y_{2k-1}^{\#} n_{2k-2} \bar{u}_{2k-2} / f_k) P_k H \sum^{k-1} n_{2m}^4 \bar{u}_{2m}^{-4} f_m \bar{B}_m / 2$$

Fifth-Order Pupil Distortion

$$[(zq)_{k-1} + (zq)_k]^{ex} = -(y_{2k-1}^{\#2} / f_k^2) \bar{E}_k \sum^{k-1} (y_{2m+1}^{\#3} n_{2m} \bar{u}_{2m} / f_m^2) \bar{E}_m / 2H$$

7.5 The Lens-Design Equations

So far in this chapter, we have defined and calculated the intrinsic and extrinsic fifth-order aberrations and have applied the results to forward and
backward modules. We have also distinguished two components of the extrinsic
contributions for a complete module; the extrinsic contribution from a surface within the module, the extrinsic internal component, and the extrinsic
contributions summed over the preceding modules, the extrinsic external component.

Our next task is to apply these results to an optical system, to obtain a
system of lens-design equations analogous to those obtained for the third-
order contributions, which appear in Tables 6.15 and 6.17. As in the case of
Table 6.15, we assume an optical system consisting of three coupled modules.

The object is at infinity, the focal length is $\mathscr{F}$, $\mathscr{B}$ the f number, and β the semi-field angle; as in (6.8.8,9). Table 6.13 contains the details of the design. Table 6.14 provides the input-output relationships and we will make use of (6.8.1-7).

In Sect.7.3, we used fifth-order spherical aberration as a model for our calculations; in Sect.7.4, we used fifth-order astigmatism for the same purpose. In this section, we will use sagittal oblique spherical aberration (SOBSA), which we have indicated as (ye) and (ze), see (7.1.2) and Table 7.4.

As in Sect.6.8, we denote the first module, which is backward, by the subscript a; the second, a forward module, by b; and the third, which is also backward, by c. We use Tables 6.13,14 to obtain data for these modules, which will be substituted into the expressions that appear in Tables 7.7-9 and Tables 7.11-13.

First, we pick up the expression for the intrinsic SOBSA contribution for a backward module, from Table 7.11

$$(ye)_a = (y_1^4 n_0^2 \bar{u}_0^2 / f_a^3)[(YE)_2^- + (YE)_1^-]_a \quad . \tag{7.5.1}$$

Recall that the numerical subscripts on $(YE)^-$ refer to the module's surfaces. Because this module is in its backward orientation, the second surface comes before the first.

Second, we refer to Table 7.12 for the extrinsic internal contribution to SOBSA from the first module,

$$[(ze)_1 + (ze)_2]^{in} = (y_1^4 n_0^2 \bar{u}_0^2 / f_a^3)(ZE)_{in\ a}^- \quad . \tag{7.5.2}$$

Next, we note that the extrinsic external contribution is zero because there are no components preceding the first module. Using the notation used in Table 7.13, we write

$$[(ze)_1 + (ze)_2]^{ex} = 0 \quad . \tag{7.5.3}$$

The second module is treated next. We use Table 7.7 for the forward module to obtain the intrinsic contribution,

$$(ye)_b = (u_2^4 n_2^4 \bar{y}_2^{\#2} / f_b)[(YE)_1^+ + (YE)_2^+]_b \quad . \tag{7.5.4}$$

From Table 6.4 for the backward orientation, we obtain

$$n_2u_2 = -y_1/f_a$$
$$\bar{y}_2 = f_a n_0 \bar{u}_0 \; , \tag{7.5.5}$$

and from (6.8.1),

$$f_b = -f_a E_a/E_b \; . \tag{7.5.6}$$

Substituting these into (7.5.4), we obtain

$$(ye)_b = -(y_1^4 n_0^2 \bar{u}_0^2 E_b/E_a f_a^3)[(YE)_1^+ + (YE)_2^+]_b \; . \tag{7.5.7}$$

The extrinsic internal contribution comes from Table 7.8, and is

$$[(ze)_3 + (ze)_4]^{in} = (u_2^4 n_2^4 \bar{y}_2^{\#2}/f_b)(ZE)_{in\;b}^+ \; . \tag{7.5.8}$$

After an application of (7.5.5,6), this becomes

$$[(ze)_3 + (ze)_4]^{in} = -(y_1^4 n_0^2 \bar{u}_0^2 E_b/E_a f_a^3)(ZE)_{in\;b}^+ \; . \tag{7.5.9}$$

The extrinsic external contribution for SOBSA is obtained from Table 7.9,

$$[(ze)_3 + (ze)_4]_b^{ex} = (u_2 n_2 \bar{y}_2^{\#}/f_b) P_b H[u_2^3 n_2^3 \bar{y}_2^{\#}(2F_a - \bar{E}_a)]/2$$
$$+ u_2^2 n_2^2 f_b [(u_2^2 n_2^2 \bar{y}_2^{\#2}/f_a)(P_a H^2 - 2\bar{c}_a)]/2H \; . \tag{7.5.10}$$

Again we apply (7.5.5,6) to get

$$[(ze)_3 + (ze)_4]_b^{ex} = -(y_1^4 n_0^2 \bar{u}_0^2 E_b/E_a f_a^3)[P_b H(2F_a - \bar{E}_a)/2 + F_b(P_a H^2 + 2\bar{c}_a)/2H] \; . \tag{7.5.11}$$

Finally, we turn our attention to the third module in this unholy trinity. We again make use of the expressions for the backward module found in Sect. 7.4. The intrinsic contribution comes first, from Table 7.11

$$(ye)_c = (y_5^{\#4} n_4^2 \bar{u}_4^2/f_c^3)[(YE)_2^- + (YE)_1^-]_c \; . \tag{7.5.12}$$

Here we make use of the fact that the first and second modules are hard-way coupled, so that

$$y_5^{\#} = y_1 E_a/E_b$$
$$n_4 \bar{u}_4 = -n_0 \bar{u}_0 E_b/E_a \quad . \tag{7.5.13}$$

We obtain these from Table 7.14.(see p.138). Moreover from (6.8.7), we have

$$f_c = \mathscr{F} E_a/E_b \quad , \tag{7.5.14}$$

where $\mathscr{F}$ represents the focal length of the entire system. We apply these to (7.5.13) to obtain

$$(ye)_c = (y_1^4 n_0^2 \bar{u}_0^2 E_b/E_a \mathscr{F}^3)[(YE)_2^- + (YE)_1^-]_c \quad . \tag{7.5.15}$$

By now the ritual should be familiar. We next do the extrinsic internal contribution from Table 7.12,

$$[(ze)_5 + (ze)_6]^{in} = (y_5^{\#4} n_4^2 \bar{u}_4^2/f_c^3)(ZE)_{in\ c}^- \quad , \tag{7.5.16}$$

which, on application of (7.5.13,14), becomes

$$[(ze)_5 + (ze)_6]^{in} = (y_1^4 n_0^2 \bar{u}_0^2 E_b/E_a \mathscr{F}^3)(ZE)_{in\ c}^- \quad . \tag{7.5.17}$$

The conclusion of this exercise consists of the calculation of the extrinsic external contribution to SOBSA. We find, from Table 7.13 and Table 7.9,

$$\begin{aligned}
[(ze)_5 + (ze)_6]^{ex} &= (y_5^{\#} n_4 \bar{u}_4/f_c) P_c H \big[(y_5^{\#3} n_4 \bar{u}_4/f_b^2)(2F_b - \bar{E}_b) + u_2^3 n_2^3 \bar{y}_2^{\#}(2F_a - \bar{E}_a) \big]/2 \\
&\quad + (y_5^{\#2}/f_c^2) F_c (y_5^{\#2} n_4^2 \bar{u}_4^2/f_b)(P_b H^2 - 2\bar{c}_b) + (u_2^2 n_2^2 \bar{y}_2^{\#2}/f_a) \\
&\quad \times (P_a H^2 - 2\bar{c}_a) \big]/2H \quad . \tag{7.5.18}
\end{aligned}$$

Upon application of (7.5.13,14), we reduce this to

$$\begin{aligned}
[(ze)_5 + (ze)_6]^{ex} &= y_1^4 n_0^2 \bar{u}_0^2 \Big\{ (E_b/E_a \mathscr{F} f_a^2) P_c H[(2F_b - \bar{E}_b) + (2F_a - \bar{E}_a)]/2 \\
&\quad + (1/\mathscr{F}^2 f_a) F_c [-(E_b/E_a)(P_2 H^2 - 2\bar{c}_2) + (P_a H^2 - 2\bar{c}_a)]/2H \Big\} \quad . \\
&\tag{7.5.19}
\end{aligned}$$

This completes the calculation of the intrinsic and extrinsic components of SOBSA. They are now summed and set equal to zero to get this lens-design equation. Adding together (7.5.1-3,7,9,11,15,17,19) and doing the obvious reductions, we obtain

$$\frac{1}{f_a^3}\left\{[(YE)_2^-+(YE)_2^-+(ZE)_{in}^-]_a-\frac{E_b}{E_a}\left([(YE)_1^++(YE)_2^++(ZE)_{in}^+]_b\right.\right.$$

$$\left.+P_bH(2F_a-\bar{E}_a)/2+F_b(P_aH^2-2\bar{C}_a)/2H)\right\}+\frac{E_b}{E_a\mathscr{F}^3}[(YE)_2^-+(YE)_1^-+(ZE)_{in}^-]_c$$

$$+\frac{E_b}{E_a\mathscr{F}f_a^2}\,P_cH[(2F_b-\bar{E}_b)+(2F_a-\bar{E}_a)]/2$$

$$+\frac{1}{\mathscr{F}f_a}F_c\left[-\frac{E_b}{E_a}(P_bH^2-2\bar{C}_b)+(P_aH^2-2\bar{C}_a)\right]/2H=0\quad. \tag{7.5.20}$$

This formidable equation is a cubic polynomial in f_a. Its solution provides a value of f_a that leads to values of the optical parameters of a lens design for which SOBSA is identically zero.

This equation, along with the other thirteen, is overwhelming. To simplify them, we introduce a somewhat less cumbersome notation. We have already reserved the earlier letters of the alphabet for the third-order aberrations. Here we will use the later letters (skipping P to avoid confusion with Petzval) of the alphabet to denote the fifth-order aberrations. We will reserve the last letter of the alphabet for seventh-order spherical aberration, which will appear in the next section. Further, we may drop the distinction between intrinsic and extrinsic internal contributions and lump them together under one symbol, indicated by the asterisk (*). As before, the plus sign (+) and the minus sign (-) signal the forward and backward orientations, respectively. These new definitions are listed in Table 7.14. Note that the quantities (YA), (YB), etc. are defined in Table 7.7 if they are accompanied by a (+); they are defined in Table 7.11 if they bear a (-). Also $(ZA)_{in}$, $(ZB)_{in}$, etc. are to be found in Table 7.8 if they have a (+) or in Table 7.12 if they have a (-). A subscript will denote the module on which the calculation is made.

Using Tables 7.14,15, we may rewrite (7.5.20) in a more convenient form,

$$(1/f_a^3)\left\{*Q_a^--(E_b/E_a)[*Q_b^++P_bH^\#Q_{1a}/2+F_b{}^\#Q_{2a}/2H]\right\}$$

$$+(E_b/E_a\mathscr{F}^3)*Q_c^-+(E_b/E_a\mathscr{F}f_a^2)P_cH(^\#Q_{1b}+{}^\#Q_{1a})/2$$

$$+(1/\mathscr{F}^2f_a)F_c[-(E_b/E_a)^\#Q_{2b}+{}^\#Q_{2a}]/2H=0\quad. \tag{7.5.21}$$

Here the remaining external contributions are indicated by the sharp sign (#).

Following exactly the same process, we can assemble the lens-design equations for all of the fifth-order aberrations listed. These are found in Table 7.16. Note here that lower-case letters as subscripts denote modules. A solution of any of the equations in Table 7.16 provides a value of f_a which assures that the appropriate fifth-order aberration will vanish identically in the three-module system.

Table 7.14. Internal contributions

Spherical Aberration	*Astigmatism*
$*K = (YA)_1 + (YA)_2 + (ZA)_{in}$	$*V = (YM)_1 + (YM)_2 + (ZM)_{in}$

Coma

Petzval

$*L = (YB)_1 + (YB)_2 + (ZB)_{in}$

$*W = (YN)_1 + (YN)_2 + (ZN)_{in}$

$*M = (YC)_1 + (YC)_2 + (ZC)_{in}$

Image Distortion

Oblique Spherical

$*X = (YP)_1 + (YP)_2 + (ZP)_{in}$

$*N = (YD)_1 + (YD)_2 + (ZD)_{in}$

$*Q = (YE)_1 + (YE)_2 + (ZE)_{in}$

Pupil Distortion

$*R = (YF)_1 + (YF)_2 + (ZF)_{in}$

$*Y = (YQ)_1 + (YQ)_2 + (ZQ)_{in}$

Elliptical Coma

$*S = (YG)_1 + (YG)_2 + (ZG)_{in}$

$*T = (YH)_1 + (YH)_2 + (ZH)_{in}$

$*U = (YK)_1 + (YK)_2 + (ZK)_{in}$

Table 7.15. External contributions

Spherical Aberration	*Oblique Spherical*
$\#K = 0$	$\#N = PH^2 + 2\bar{C}$

Coma	$\#Q_1 = 2F - \bar{E}$
$\#L = 5F - 4\bar{E}$	$\#Q_2 = PH^2 - 2\bar{C}$
$\#M = 2F - \bar{E}$	$\#R_1 = \bar{C}$
	$\#R_2 = F$

Astigmatism	*Elliptical Coma*
$\#V_1 = PH^2$	$\#S_1 = F$
$\#V_2 = \bar{F}$	$\#S_2 = PH^2 + \bar{C}$
$\#V_3 = \bar{B}$	$\#S_3 = E - 2\bar{F}$
	$\#T_1 = F$
Petzval	$\#T_2 = PH^2 - \bar{C}$
$\#W_1 = PH^2$	$\#T_3 = E - 8\bar{F}$
$\#W_2 = 4E - \bar{F}$	$\#U_1 = F$
$\#W_3 = \bar{B}$	$\#U_2 = PH^2 - \bar{C}$
	$\#U_3 = E - 4\bar{F}$
Image Distortion	
$\#X_1 = E$	
$\#X_2 = \bar{B}$	

Pupil Distortion

$\#Y = \bar{E}$

Table 7.16. The lens-design equations

Spherical Aberration

$$(1/f_a^5)[*K_a^- - (E_a/E_b)*K_b^+] + (E_a/E_b\mathscr{F}^5)*K_c^- = 0$$

Coma

$$(1/f_a^4)[*L_a^- - *L_b^+ - F_b\#L_a/2H] - (1/\mathscr{F}^4)*L_c^- + (E_a/E_b\mathscr{F}^2)F_c[\#L_b/{}^2 - \#L_a/f_a^2]/2H = 0$$

$$(1/f_a^2)[*M_a^- - *M_b^+ - F_b\#M_a/H] - (1/\mathscr{F}^4)*L_c^- + (1/f_a^2\mathscr{F}^2)F_c(\#M_b + \#M_a)/H = 0$$

Oblique Spherical

$$(1/f_a^3)[*N_a^- - (E_b/E_a)*N_b^+ + \#N_a/H] + (1/\mathscr{F}^3)*N_c^- + (1/f_a\mathscr{F}^2)[(E_b/E_a)\#N_b + \#N_a]/H = 0$$

$$(1/f_a^3)\{*Q_a^- - (E_b/E_a)[*Q_b^+ + P_bHQ_{1a}/2 + F_b^\#Q_{2a}/2H]\} + (E_b/E_a\mathscr{F}^3)*Q_c^- + (E_b/E_a\mathscr{F}^2)P_cH(\#Q_{1b} + \#Q_{1a})/2$$
$$+ (1/\mathscr{F}^2 f_a)F_c[-(E_b/E_a)\#Q_{2b} + \#Q_{2a}]/2H = 0$$

$$(1/f_a^3)[*R_a^- - (E_b/E_a)*R_b^+ - 2(E_a/E_b)P_bH\#R_{2a} + F_b\#R_{1a}/H] + (1/\mathscr{F}^3)*R_c^- + (2/\mathscr{F}^2 f_a)F_c[(E_b/E_a)R_{1b} + R_{1a}]$$
$$+ (2E_b/E_a\mathscr{F} f_a^2)P_cH(\#R_{2b} + \#R_{1b}) = 0$$

Elliptical Coma

$$(1/f_a^2)[*S_a^- - (E_b/E_a)^2*S_b^+ - 3(E_b/E_a)^2 E_b\#S_{1a}/2H - (E_b/E_a)P_bH\#S_{2a}/2 + F_b\#S_{3a}/2H] - (E_a^2/E_b^2\mathscr{F}^2)*S_c^- + 3(E_b^2/E_a^2 f_a^2)E_c(\#S_{1b} + \#S_{1a})/2H$$
$$+ (E_b/E_a\mathscr{F} f_a)P_cH[-(E_b/E_a)\#S_{2b} + \#S_{2a}]/2 - (1/\mathscr{F}^2)F_c[(E_b/E_a)^2\#S_{3b} + \#S_{3a}]/2H = 0$$

$$(1/f_a^2)[*T_a^- - (E_b/E_a)^2*T_b^+ + E_b\#T_{1a}/2H - (E_b/E_a)P_bH\#T_{2a}/2 + (E_b/E_a)^2 F_b\#T_{3a}/2H] - (E_b^2/E_a^2\mathscr{F}^2)*T_c^- + 3(E_b^2/E_a^2 f_a^2)E_c(\#T_{1b} + \#T_{1a})/H$$
$$+ (E_b/E_a f_a\mathscr{F})P_cH[-(E_b/E_a)\#T_{2b} + \#T_{2a}] + (1/\mathscr{F}^2)F_c[(E_b/E_a)^2\#T_{3b} + \#T_{3a}]/H = 0$$

$$(1/f_a^2)[*u_a^- + (E_b/E_a)^2*u_b^+ + (E_b/E_a)^2 E_b\#u_{1a}/2H - (E_b/E_a)P_bH\#u_{2a}/2 + F_b\#u_{3a}/2H] - (E_b^2/E_a^2\mathscr{F}^2)*u_c^- + (E_b^2/E_a^2 f_a^2)E_c(\#u_{1b} + \#u_{1a})/2$$
$$- (E_b/E_a f_a\mathscr{F})P_cH[(E_b/E_a)\#u_{2b} + \#u_{2a}]/2 + (1/\mathscr{F}^2)F_c[(E_b/E_a)^2\#u_{3b} + \#u_{3a}]/2H = 0$$

Table 7.16 (cont.)

Astigmatism

$(1/f_a)[*v_a^- - (E_b/E_a)^3*v_b^+ + E_b^\#v_{1a}/4H - (E_b/E_a)P_b H\#v_{2a} - f_b\#v_{3a}/2H] + (E_b/E_a)^3*v_c^- + (E_b^2/E_a^2 f_a)E_c[(E_b/E_a)\#v_{1b} + \#v_{1a}]/4H$

$+ (f_z/\mathscr{F}^2)F_c[-(E_b/E_a)^4\#v_{3b} + \#v_{3a}]/2H + (1/\mathscr{F})P_c H[(E_b/E_a)^3\#v_{2b} + \#v_{2a}]/2 = 0$

Petzval

$(1/f_a)[*w_a^- - (E_b/E_a)^3*w_b^+ + (E_b/E_a)^2 E_b\#w_{1a}/2H + (E_b/E_a)P_b H\#w_{2a}/2 - f_b\#w_{3a}/H] + (E_b^3/E_a^3\mathscr{F})*w_c^-$

$+ (E_b^2/E_a^2 f_a)E_c[-(E_b/E_a)\#w_{1b} + \#w_{1a}]/2H + (E_b/E_a^2\mathscr{F})P_c H[(E_b/E_a)^2\#w_{2b} + \#w_{2a}]/2 + (f_a/\mathscr{F}^2)F_c[-(E_b/E_a)\#w_{3b} + \#w_{3a}]/H = 0$

Image Distortion

$*X_a^- - (E_b/E_a)^4*X_b^+ - (E_b/E_a)^2 F_b\#X_{1a}/2H - (E_b/E_a)P_b H\#X_{2a}/2 - (E_b/E_a)^4*X_c^-$

$+ (E_b/E_a)^2 E_c[(E_b/E_a)^2\#X_{1b} + \#X_{1a}]/2H + (F_a/F)P_c H[(E_b/E_a)^3\#X_{2b} + \#X_{2a}] = 0$

Pupil Distortion

$(1/f_a^4)[*v_a^- - *v_b^+ + \bar{E}_b\#v_a/2H] - (1/\mathscr{F}^4)*v_c^- + (1/\mathscr{F}^2 f_a^2)\bar{E}_c[\#v_a + \#v_b]/2H = 0$

7.6. Seventh-Order Spherical Aberration

The final section of this chapter on fifth-order aberrations is concerned with the calculation of seventh-order spherical aberration. This may appear a little strange. However, because we are dealing with seventh-order *spherical* aberration, its calculation follows the pattern we have become familiar with in the fifth-order calculations. Indeed, we shall follow these patterns slavishly in the subsequent calculations.

The starting point is RIMMER's formula for the intrinsic contribution

$$\tilde{B}_7 = Si^2[10\omega^2 + Sic(2u - 5u_{-1}/8n_{-1}] \ . \tag{7.6.1}$$

In our notation, defined in Tables 7.1-3, this becomes

$$(yz)_j = y_j s_j n_j (a_j/n_{j-1})^2 [10W_j + y_j s_j n_j a_j s_j (2u_j - 5u_{j-1})/8N_{j-1}^2] \ . \tag{7.6.2}$$

Using the procedures developed in Sects.7.3,4, we obtain the forward and backward canonical versions. These are, for the forward module,

$$(yz)_j = u_{j-1}^8 n_{j-1}^8 f(YZ)_1^+ \ , \tag{7.6.3}$$

where

$$(YZ)_1^+ = Y_1 S_1 N_1 (A_1/N_0)^2 \left[10W_1^{+2} + Y_1 S_1 N_1 A_1 C_1 (2U_1 - 5U_0)/8N_0^2 \right] \ . \tag{7.6.4}$$

Equation (7.6.4) is the canonical form for the first surface of a module. A similar expression is obtained for the second surface.

For the backward module, we get

$$(yz)_j = y_j^{\#8}/f^7)(YZ)_2^- \ , \tag{7.6.5}$$

where

$$(YZ)_2^- = Y_2 S_2 N_2 (A_2/N_2)^2 \left[10W_2^{-2} + Y_2 S_2 N_2 A_2 C_2 (2U_1 - 5U_2)/8N_2^2 \right] \ . \tag{7.6.6}$$

Equation (7.6.6) is the canonical version for the contribution at the first surface of the backward module. A similar expression can be derived in an obvious way for the second-surface contribution.

We next redefine these quantities to obtain a form suitable for the lens-design equations and which is consistent with the definitions in Table 7.14. For the forward orientation,

$$[(yz)_j + (yz)_{j+1}] = u'^7_{j-1} n'^7_{j-1} f[(YZ)^+_1 = (YZ)^+_2] \quad . \tag{7.6.7}$$

For the backward orientation, we use

$$[(yz)_j = (yz)_{j+1}] = (y^{\#8}_j / f^7) [(YZ)^-_2 + (YZ)^-_1] \quad . \tag{7.6.8}$$

Next come the extrinsic contribution. We begin with RIMMER's formula,

$$^0B_7 = (1/2I)[(1/2I)(n+3c)B'^2 + 3F(B'_5 - B'\bar{E}'/I) + 3B(\bar{E}'^2/2I - \bar{E}'_5 + (\tilde{F}_1 + \tilde{F}_2)B' - 5B_5\bar{E}'] \quad . \tag{7.6.9}$$

Translating this into the notation that we are using, (refer to Tables 7.1-4), we obtain

$$(ZZ)_j = (1/2h)^2 \Big[(P_j h^2 + 3c_j) \Big(\sum^{j-1} b_k \Big)^2 - 6f_j \Big(\sum^{j-1} b_k \Big) \Big(\sum^{j-1} \bar{e}_k \Big)$$
$$+ 3b_j \Big(\sum^{j-1} \bar{e}_k \Big)^2 \Big] + (1/2h) \Big\{ 3f_j \sum^{j-1} [(ya)_k + (za)_k] - 3b_j \sum^{j-1} [(yq)_k + (zq)_k]$$
$$+ [(yb)_j + (yc)_j] \sum^{j-1} b_k - 5(ya)_j \sum^{j-1} \bar{e}_k \Big\} \quad . \tag{7.6.10}$$

The next step in this rather intricate operation is to sum the extrinsic contributions from the two surfaces of a module and then to separate the result into its internal and external parts. As before, we make use of the fact that third-order spherical aberration and third-order astigmatism are zero over a module so that, for the kth module in an optical system

$$b_{2k-1} + b_{2k} = 0 \quad , \quad \sum^{k-1} (b_{2m-1} + b_{2m}) = 0 \tag{7.6.11}$$

$$c_{2k-1} + c_{2k} = 0 \quad , \quad \sum^{k-1} (c_{2m-1} + c_{2m}) = 0 \quad . \tag{7.6.12}$$

With these identities, we obtain from (7.6.10)

$$(zz)_{2k-1} + (zz)_{2k} = (1/2h)^2 (P_{2k}h^2 + 3C_{2k})b_{2k-1}^2 - 6\delta_{2k}b_{2k-1} \sum^{k-1} (\bar{e}_{2m} + \bar{e}_{2m-1})$$

$$- 6\delta_{2k}b_{2k-1}\bar{e}_{2k-1} + 6b_{2k}\bar{e}_{2k-1} \sum^{k-1} (\bar{e}_{2m} + \bar{e}_{2m-1}) + 3b_{2k}\bar{e}_{2k-1}^2$$

$$+ (1/2h)\Big\{ 3(\delta_{2k} + \delta_{2k-1}) \sum^{k-1} [(ya)_{2m} + (ya)_{2m-1} + (za)_{2m} + (za)_{2m-1}]$$

$$+ 3\delta_{2k}[(ya)_{2k-1} + (za)_{2k-1}] - 3b_{2k}[(yq)_{2k-1} + (zq)_{2k-1}]$$

$$+ [(yb)_{2k} + (yc)_{2k}]b_{2k-1} - 5[(ya)_{2k} + (ya)_{2k-1}] \sum^{k-1} (\bar{e}_{2m} + \bar{e}_{2m-1})$$

$$- 5(ya)_{2k}\bar{e}_{2k-1} \Big\} \ . \tag{7.6.13}$$

Equation (7.6.13) is next separated into its internal and external parts,

$$[(zz)_{2k-1} + (zz)_{2k}]^{in} = (1/2h)^2 \Big[(P_{2k}h^2 + 3C_{2k})b_{2k-1}^2 - 6\delta_{2k}b_{2k-1}\bar{e}_{2k-1}$$

$$+ 3b_{2k}\bar{e}_{2k-1}^2 \Big] + (1/2h)\Big\{ 3f_{2k}[(ya)_{2k-1} + (za)_{2k-1}]^{in}$$

$$- 3b_{2k}[(yq)_{2k-1} + (zq)_{2k-1}]^{in} + [(yb)_{2k} + (yc)_{2k}]^{in}b_{2k-1}$$

$$- 5(ya)_{2k}\bar{e}_{2k-1} \Big\} \ , \tag{7.6.14}$$

$$[(zz)_{2k-1} + (zz)_{2k}]^{ex}$$

$$= (-1/2h^2)\Big\{ 3\delta_{2k}b_{2k-1} - 3b_{2k}\bar{e}_{2k-1} + 5h[(ya)_{2k} + (ya)_{2k-1}] \Big\} \sum^{k-1} (\bar{e}_{2m} + \bar{e}_{2m-1})$$

$$+ (3/2h)\Big\{ (f_{2k} + f_{2k+1}) \sum^{k-1} [(ya)_{2m} + (ya)_{2m-1} + (za)_{2m} + (za)_{2m-1}]$$

$$+ \delta_{2k}(za)_{2k-1}^{ex} - b_{2k}(zq)_{2k-1}^{ex} \Big\} \ . \tag{7.6.15}$$

Now we come to the calculation of the canonical versions of the extrinsic internal contributions given in (7.6.14) for the forward and backward cases. We do the forward case first, using the equations in Tables 7.1-4, and obtain, after the usual struggle,

$$[(zz)_{2k-1} + (zz)_{2k}]^{in} = u_{2k-2}^8 n_{2k-2}^8 f_k \Big\{ (1/2H)^2 [(P_2H^2 + 3C_2)B_1 - 6F_2B_1\bar{E}_1$$

$$+ 3B_2\bar{E}_1^2] + (1/2H)\Big[3F_2(YA)_1^+ - 3B_2(YQ)_1^+$$

$$+ B_1[(YB)_2^+ + (YC)_2^+] - 5\bar{E}_1(YA)_2^+ \Big] \Big\} \ . \tag{7.6.16}$$

In these last calculations, note that $(za)^{in}_{2k-1}$ and $(zq)^{in}_{2k-1}$ are both zero. There is no extrinsic internal contribution at the first surface of a module. The backward orientation follows from (7.6.14) in pretty much the same way,

$$[(zz)_{2k-1}+(zz)_{2k}]^{in} = (\bar{y}^{\#8}_{2k-1}/f^7_k)\{(1/2H)^2[(P_1H^2+3C_1)B^2_2-6F_1B_2\bar{E}_2+3B_1\bar{E}^2_2]$$

$$+(1/2H)[3F_1(YA)^-_1-3B_1(YQ)^-_2+B_2[(YB)^-_1+(YC)^-_1]-5\bar{E}_2(YA)^-_1]\}.$$

$$(7.6.17)$$

The next step is to calculate the canonical versions of the extrinsic external contributions for the forward and backward orientations. The forward orientation is treated first; use is made of (7.6.16). It is most convenient to do this in three separate steps. The first term in (7.6.16) is

$$(1/2h)^2\{3\textit{b}_{2k}b_{2k-1}-3b_{2k}\bar{e}_{2k-1}+5h[(ya)_{2k}+(ya)_{2k-1}]\}\sum^{k-1}(\bar{e}_{2m}+\bar{e}_{2m-1})$$

$$= (u^5_{2k-2}n^5_{2k-2}f_k/\bar{y}^\#_{2k-1})(1/2H)^2[3B_2\bar{E}_1-3F_2B_1-5H((YA)^+_1+(YA)^+_2)]_k \sum^{k-1} u^3_{2m}n^3_{2m}\bar{y}^\#_{2m}\bar{E}_m \cdot$$

Here $\bar{E}_m = \bar{e}_{2m} + \bar{e}_{2m-1}$, which represents the sum of pupil distortion over the mth module. Before calculating the second term, note that, from Table 7.9, $(za)^{ex} = 0$, so that $(za)_{2m} + (za)_{2m-1} = [(za)_{2m}+(za)_{2m-1}]^{in} = u^6_{2m}n^6_{2m}f_m[(ZA)^+_{in}]_m$. This last expression is found in Table 7.8. Also, note from Table 7.9 that

$$(zq)^{ex} = -u^2_{2k-2}n^2_{2k-2}\bar{E}_k \sum^{k-1} u^3_{2m}n^3_{2m}\bar{y}^\#_{2m}\bar{E}_m/2H \cdot$$

When these results are used in the second term of (7.6.16), we obtain

$$(3/2h)\{(f_{2k}+f_{2k+1})\sum^{k-1}[(ya)_{2m}+(ya)_{2m-1}+(za)_{2m}+(za)_{2m-1}]$$

$$+f_{2k}(za)^{ex}_{2k-1}-b_{2k}(zq)^{ex}_{2k-1}\}$$

$$+u^2_{2k-2}n^2_{2k-1}\{(3/2H)F_k \sum^{k-1} u^6_{2m}n^6_{2m}f_m[(YA)^+_1+(YA)^+_2+(ZA)^+_{in}]_m$$

$$-3(u^5_{2k-2}n^5_{2k-2}f_k/\bar{y}^\#_{2k-2})(B_2)_k\bar{E}_k \sum^{k-1} u^3_{2m}n^3_{2m}\bar{y}^\#_{2m}\bar{E}_m\} / 2H \cdot$$

Now, putting these two terms together, we obtain

$$[(zz)_{2k-1}+(zz)_{2k}]^{ex} = (u^5_{2k-2}n^5_{2k-2}f_k/\bar{y}^\#_{2k-1})[(1/2H)^2\{3B_2\bar{E}_1-3F_2B_1$$

$$-5H[(YA)^+_1+(YA)^+_2]-3B_2\bar{E}\}_k \sum^{k-1} u^3_{2m}n^3_{2m}\bar{y}^\#_{2m}\bar{E}_m]$$

$$+u^2_{2k-2}n^2_{2k-2}[(3/2H)F_k \sum^{k-1} u^6_{2m}n^6_{2m}f_m*K_m] \cdot \qquad (7.6.18)$$

We do the same sort of thing for the backward orientation and obtain

$$[(zz)_{2k-1} + (zz)_{2k}]^{ex} = (\bar{y}_{2k-1}^{\#5}/n_{2k-2}\bar{u}_{2k-2}f_k^5)\Big((1/2H)^2\Big\{3B_1\bar{E}_2 - 3F_1B_2$$

$$-5H[(YA)_2^- + (YA)_1^-] - 3B_1\bar{E}\Big\}_k \sum^{k-1}(y_{2m+1}^{\#3}\bar{u}_{2m}n_{2m}/f_m^2)\bar{E}_m\Big) \quad .$$

(7.6.19)

Before going on to the lens-design equations, let us define some more compact canonical relations. These follow the general pattern established in Tables 7.14,15. We use data from (7.6.16-19),

$$*Z^+ = (YZ)_1^+ + (YZ)_2^+ + (1/2H)^2[(P_2H^2 + 3C_2)B_1 - 6F_2B_1\bar{E}_1 + 3B_2\bar{E}_1^2]$$

$$+ (1/2H)\Big\{3F_2(YA)_1^+ - 3B_2(YQ)_1^+ + B_1[(YB)_2^+ + (YC)_2^+] - 5\bar{E}_1(YA)_2^+\Big\} \quad , \qquad (7.6.20)$$

$$*Z^- = (YZ)_2^- + (YZ)_2^- + (1/2H)^2[(P_1H^2 + 3C_2)B_1 - 6F_1B_2\bar{E}_2 + 3B_1\bar{E}_2^2]$$

$$+ (1/2H)\Big\{3F_1(YA)_2^- - 3B_1(YQ)_2^- + B_2[(YB)_1^- + (YC)_1^-] - 5\bar{E}_2(YA)_1^-\Big\} \quad , \qquad (7.6.21)$$

$$\#Z_1^+ = (1/2H)^2\Big\{3B_2\bar{E}_1 - 3F_2B_1 - 5H[(YA)_1^+ + (YA)_2^+] - 3B_2\bar{E}\Big\} \quad , \qquad (7.6.22)$$

$$\#Z_1^- = (1/2H)^2\Big\{3B_1\bar{E}_2 - 3F_1B_2 - 5H[(YA)_2^- + (YA)_1^-] - 3B_1\bar{E}\Big\} \quad , \qquad (7.6.23)$$

$$\#Z_2^- = 3F/2H \quad . \qquad (7.6.24)$$

Finally, we consider the three-module system described in Sect.7.5 and obtain the lens-design equation for seventh-order spherical aberration. From (7.6.9,18,20), we obtain expressions for the internal and external contributions for the first module,

$$(YZ)_a + (ZZ)_a^{in} = (y_1^{\#8}/f_a^7)*Z_a^- \quad ,$$

$$(zz)_a^{ex} = 0 \quad . \qquad (7.6.25)$$

For the second and third modules, we use the relations found in Sects.6.6 and 7.5 and obtain,

$$(YZ)_b(ZZ)_b^{in} = (y_1^{\#8}/f_a^7)(E_a/E_b)*Z_b^+$$

$$(zz)_b^{ex} = (y_1^{\#8}/f_a^7)\Big\{-(E_a/E_b)^{\#}Z_{1b}^+\bar{E}_a + {}^{\#}Z_{2b}[(YA)_1^+ + (YA)_2^+]_a\Big\} \quad . \qquad (7.6.26)$$

For the third module,

$$(YZ)_c + (ZZ)_c^{in} = (y_1^{\#8}/\mathscr{F}^7)(E_a/E_b)*Z_c^-$$

$$(ZZ)_c^{ex} = (y_1^8/\mathscr{F}^5 f_a^2)(E_a/E_b)\#Z_{1c}^-(\bar{E}_b+\bar{E}_a)+(y_1^8/\mathscr{F}^2 f_a^5)\#Z_{2c}$$

$$\cdot\left\{(E_a/E_b)[(YA)_2^-+(YA)_1^-]_b+[(YA)_1^++(YA)_2^+]_c\right\} \quad . \tag{7.6.27}$$

At long last, we assemble the lens-design equation for seventh-order spherical aberration,

$$(1/f_a^7)\left\{*Z_a^-+(E_a/E_b)*Z_b^+-(E_a/E_b)\#Z_{1b}^+\bar{E}_a+\#Z_{2b}[(YA)_1^++(YA)_2^+]_a\right\}$$

$$+ (1/f_a^5\mathscr{F}^2)\#Z_{2c}\left\{(E_a/E_b)[(YA)_2^-+(YA)_1^-]_b+[(YA)_1^++(YA)_2^+]_c\right\}$$

$$+ (1/f_a^2\mathscr{F}^5)(E_a/E_b)\#Z_{1c}^-(\bar{E}_b+\bar{E}_a)+(1/\mathscr{F}^7)(E_a/E_b)*Z_c^- = 0 \quad . \tag{7.6.28}$$

This is an equation of degree seven in $1/f_a$. Its solution will yield values of f_a that will lead to a system with zero seventh-order spherical aberration.

8. Some Examples

It was my intention to include in this chapter some sort of comprehensive
study of the process of modular optical design. Time and fortune have con-
spired against me. Instead, I include a selection of some of our experiments,
hoping that these will aid the reader and potential user in formulating his
own ideas of the system's capabilities and deficiencies.

Most of the examples presented are taken from the work of MERCADO and
ANDERSON; they include some material that has already appeared in the literature. However, much of this, although not new, was originally prepared for
only internal consumption and distribution. Sources are cited when they are
known.

8.1 Camera Lens

This title is inaccurate. Here we will consider the class of optical systems
that image an object at infinity onto a finite plane. We will require that
this image be real. This excludes copy lenses, which are designed for a pair
of finite conjugates, but are also camera lenses. Here I will emulate a well-
known political figure and assert that any term I use means what I want it
to mean at the time I use it, previous and subsequent usage notwithstanding.

So, the kind of lens to be considered is one that will form an image of a
distant object on a film plane. Parameters descriptive of such lenses include
focal length, f number and semifield angle. In these examples, we will assign
values to these quantities *a priori*.

Clearly, the realization of a camera lens in terms of modules will be constrained by the following considerations. 1) The number of modules in the
design must be odd. 2) To assure a real image, the last module must be of
either the A or the B category. It turns out that this requires that the

penultimate medium must not be air. 4) Two contiguous media must not have the same refractive index. 5) The coupling conditions must be rigorously adhered to.

Initially, we looked at three-module systems. The shape parameters were adjusted so that each of the modules was of an appropriate category and satisfied the necessary conditions for coupling. It was often necessary to make further minor adjustments to satisfy sufficient conditions on the hard-way couples. We may go further, and adjust the parameters to correct third-order distortion or primary lateral chromatic aberration, or both.

The power parameter of the last module is determined by the focal length of the desired lens. The power parameters of the first two modules are connected by the conditions for hard-way coupling. At this point, we may introduce one of the lens-design equations and solve it for the power parameter of the first module. So, with a three-module system, we may obtain a design for which one of the third- or fifth-order aberrations is zero. This always works, in principle. In practice, this may lead to absurd systems in which thicknesses are negative or curvatures impossibly large. Then, further adjustments must be made of the shape parameters or the refractive indices. ANDERSON [Ref.8.1, pp.108-121] has studied relationships between these parameters that lead to the exclusion of certain values, in certain circumstances.

The design of a five-module system proceeds in exactly the same way, the difference being that two lens-design equations may be solved simultaneously. With a seven-module system all remaining aberrations may be made equal to zero. The equation for zero coma, zero Petzval and zero axial color can, in general, be solved simultaneously. Or, alternatively, a mixture of third- and fifth-order aberrations can be made zero at the same time.

There is the matter of the starting point. ANDERSON suggested that an initial value of the shape parameter should be near ψ^*. According to his argument, a problem in modular designs is that they tend to be bizarre; long things with thick glass elements or short ones that are impossibly thin. ANDERSON argued for an *ad hoc* approach by choosing values of the shape parameters which produce values of T_1 that are less than $1/C_1$ or $1/C_2$. This is particularly desirable if T_1 lies in a glass medium. In general this condition is most likely to be satisfied in a neighborhood of ψ^*.

This criterion does help in the selection of glasses. Each module must be in an appropriate category for coupling for values of the shape parameter in the neighborhood of ψ^*. This places conditions on the choice of glasses.

It turns out that ψ_c, the value of the shape parameter corresponding to a concentric lens, invariably lies near ψ^*. Thus the modules in ANDERSON's de-

signs tend to be *almost* concentric, resulting in an additional favorable dis-
position of the asymmetric aberrations.

MERCADO, on the other hand, tended toward selecting initial values in the
outer limits of the ranges of the shape parameters. He prefered the region
in which the Petzval contribution vanishes, as a starting point. He then pro-
ceeded in very much the same way as ANDERSON. His results tend to favor flat
fields, whereas ANDERSON's designs tend to better control of the asymmetric
aberrations.

A third approach, one which, in my impatience, I prefer, is to flip a coin.

Indeed, we did rely on stochastic decision-making on occasion, especially
in the earliest stages of our experiments. We wanted to see what would happen
in the very worst case, when the selection of the glasses that constitute a
system was made without any preconceived idea of proper ordering. Twenty five
glasses were chosen at random from the Schott catalog; their characteristics
were recorded on file cards, which were then shuffled and dealt. Several jokers
representing air spaces, were included in the deal. The order in which the
glasses and air spaces appeared determined the composition of the lens. Of
course, any obviously impossible sequence, such as a pair of jokers in suc-
cession resulted in a misdeal.

We felt at the time that this procedure of selecting the glasses at random
would provide the most stringent test of the system. It passed the test easily.
Under these conditions, we were able to produce designs that possessed zero
spherical, zero astigmatism, and zero whatever other aberrations were specified.

However, this random selection of glasses led to some peculiar results. One
of our earliest products was a series of designs based on seven modules, which
we fondly called the HYPOTHETIGON. One of these monsters is shown in Fig.8.1
and in Table 8.1 (see pp.158-159). In retrospect, it is hard to see why we
were so elated about it. Perhaps it is the kind of pride of parenthood that
would cause, say, a Dr. Frankenstein to dote on a most peculiar offspring.
However, Dr. Mercado, not Dr. Frankenstein, is the actual parent.

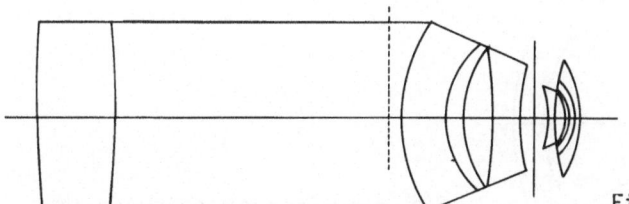

Fig. 8.1. The HYPOTHETIGON

At any rate, the HYPOTHETIGON did show that we were on the right track. Note that all of the listed aberrations, except distortion and lateral color, are zero. This was true for all versions of the HYPOTHETIGON.

Note that the front end of the lens contains several glasses whose refractive indices are rather close to each other. This evidently led to the rather massive pieces of glass at that end. It was a result of the random selection of glasses, as was the almost complete absence of air spaces. This was a case where there were not enough jokers in the deck.

A less bizarre design is shown in Fig.8.2 and Table 8.2 (see pp.159-161). The overall configuration remained the same, except that certain of the glasses were replaced by air spaces. However, it was necessary to force the coupling between the forth and fifth modules, in order to keep the stop from becoming virtual. The price that was paid was the loss of one of the power parameters; consequently, one of the aberrations could not be set to zero. Longitudinal color was the sacrificial lamb.

The next three designs are concerned with the correction of fifth-order aberrations. These were done by ANDERSON [8.2]. Two of these are shown in

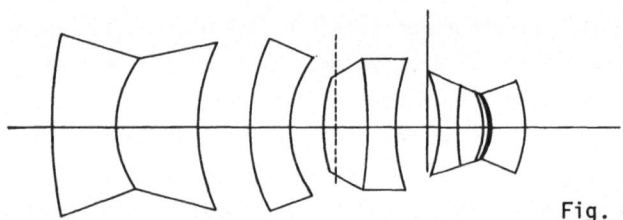

Fig. 8.2. A seven-module system

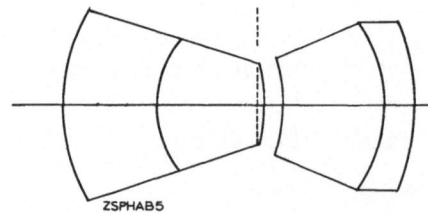

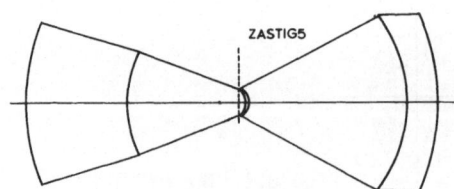

Fig. 8.3. Two three-module systems

Fig.8.3. (The third is too large to fit on the page, if drawn to the same scale.) Each of these designs employed the same three modules. Moreover, the shape parameters have exactly the same value in all three. This is to illustrate the fact that the only difference in these three lenses is in the power parameters of the first two modules. The power parameter of the third module is determined by the prescribed focal length of the lens; it, too, remains constant. The three modules are described in Table 8.3 (see p.161).

The first of the three-module designs shown in Fig.8.3, defined in Table 8.4 (see p.162), has fifth-order spherical aberration set equal to zero, as well as third-order spherical aberration and astigmatism. The prescription of the second of these lenses is found in Table 8.5 (see p.163). It has zero TOBSA. The third member of the party shown in Table 8.6 (see p.164), is set for zero fifth-order astigmatism. Note that, in this series, the values of third-order distortion and primary lateral color remain exactly the same throughout the experiment.

The next lens was developed out of the preceding three. The shape parameters were adjusted to make third-order distortion and third-order pupil coma vanish simultaneously. It is shown in Fig.8.4 and Table 8.7 (see p.165). It has zero TOBSA.

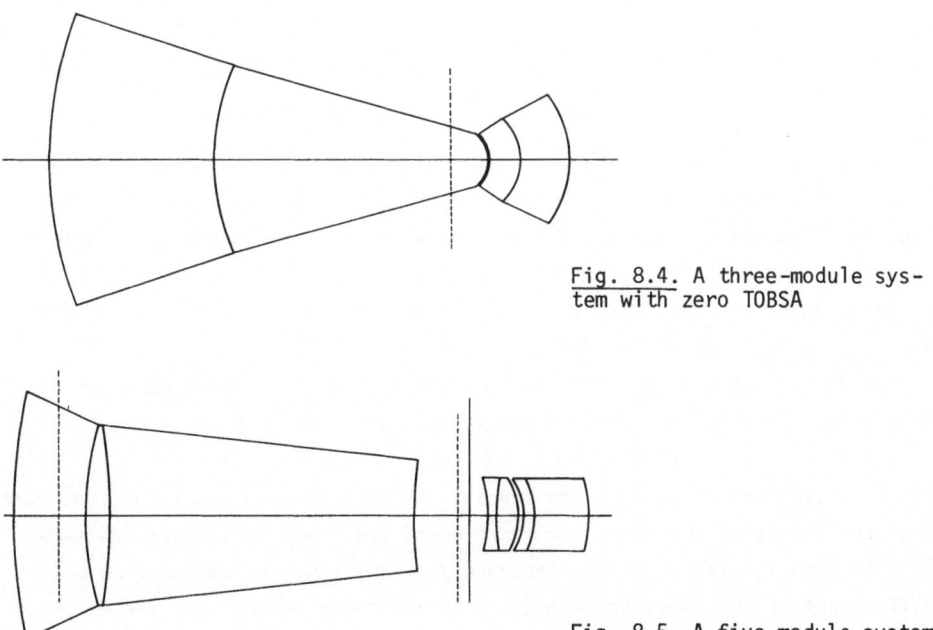

Fig. 8.4. A three-module system with zero TOBSA

Fig. 8.5. A five-module system

Figure 8.5 and Table 8.8 (see pp.166-167) show a five-module system, one of a series of about twenty five. In this series, the Petzval sum and third-order coma were set equal to zero. The shape parameters were adjusted so that both third-order distortion and primary lateral chromatic aberration were zero.

Then the shape parameters were altered in such a way that distortion and lateral color were kept equal to zero. This was done by means of an iterative routine operating on two of the shape parameters. The procedure was as follows. An arbitrary change was made in one of the shape parameters. The change was not large enough to change the category of the module. Then the routine was applied to the shape parameters of two modules different from the first, which brought the distortion and lateral color to zero values simultaneously. Then the lens-design equations for zero Petzval and coma were solved.

The design shown here was one for which primary axial chromatic aberration was at a minimum. This study was also done by ANDERSON.

8.2   The Method

The best way to describe the process of designing modules is to demonstrate. Tables 8.9-11 (see pp.167-169) show printouts of a computer program that calculates critical values and a table of canonical parameters. In each of the three runs shown, a real-case parameter value is selected and it and the values of the canonical parameters are stored in the three files; ONE%, TWO%, and THREE%.

Table 8.12 (see pp.170-171) shows a printout of another program which combines the three modules represented by files ONE%, TWO%, and THREE% and combines them into an optical design. Because this is a three-module system, we may correct only one of the three outstanding aberrations, Petzval, Coma, or Axial color. In this run, we do all three, one after the other, and store the results in files ZPETZ, ZCOMA, and ZAXIAL.

Tables 8.13-15 (see pp.172-174) show the results of a third computer program that traces paraxial rays and calculates the third-order aberrations.

One thing should be noted here. The shape parameters were chosen to make lateral color add up to zero. No precautions were taken with distortion. Note that the values of third-order distortion of the three lenses are the same. This is because distortion is independent of the power parameter. The only differences of the constituent modules in the three lenses, are the values of the three power parameters.

8.3 Copy Lenses

This title is another misnomer. What we mean here is an optical system designed to image one plane located relatively close to a lens onto another
similarly situated. The arrangement calls for an even number of modules, the
first of which is in a forward orientation and the last, in a backward orientation.

A lens of this type can be visualized as a pair of camera lenses set back-
to-back. The first camera lens images the finite object plane at infinity,
which is then imaged by the second camera lens onto its image plane. A datum
describing this type of lens is its magnification. It is easy to see that
this is the ratio of the focal lengths of the two lenses.

In the context of modular optical design, this specification of a magnification affects only the ratio of the power parameters of the first and the
last module. Other first-order parameters that can be used are equivalent
focal length, effective f number, and format size. Front and rear focal distances can also be specified. MERCADO [Ref.8.3, pp.171-175] derived relations
for these quantities and equivalent focal length in terms of paraxial quantities that can then be reduced to canonical expressions.

The procedure is very similar to that used in the camera lens. First, the
shape parameter is adjusted to assure proper coupling and, perhaps, also to
correct distortion and lateral color. Then, the magnification and one other
dimension, equivalent focal length or format size, determine the power parameters of the first and last modules. If there are 2n modules in the system,
there are n-1 pairs of hard-way-coupled modules in between, and therefore the
possibility of setting to zero n-1 aberrations.

Figure 8.6 and Table 8.16 (see pp.175-177) from MERCADO [Ref.8.3, pp.128-130]
give a six-module system. The magnification and object format size were set
beforehand, which fix the first and last power parameter. This left two power
parameters free for zeroing aberrations. Third-order coma and the Petzval sum

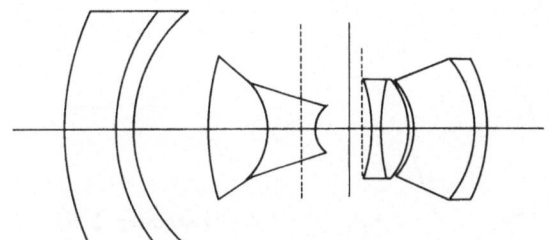

Fig. 8.6. A six-module copy
lens

154

were selected. In addition, the shape parameters were adjusted to make distortion equal to zero.

The next several designs are based on an observation of MERCADO [Ref.8.3,
pp.132-138]. Consider a symmetrical optical system. This is an optical system
with an even number of surfaces, with a stop located exactly half-way between
the two central surfaces. In addition, the first half is an exact mirror image
of the second half. That is to say the curvature of the first surface is equal
in magnitude and opposite in sign to the last. Similarly, the separation between the first two surfaces and the last two are also equal. In such a system, coma, distortion, and lateral color are identically zero.

Now consider a symmetric lens that is also a modular design. The first and
last modules will be, except for orientation, identical. This includes both
the power parameter and the shape parameter. This will be true also for every
other symmetric pair of modules in the design. Now suppose that we change the
values of the power parameters in such a way that the design modules remain
properly coupled. Because the values of the individual shape parameters remain unaltered and because distortion and lateral color depend on only these,
the value of these two aberrations will remain unchanged and only coma will
vary.

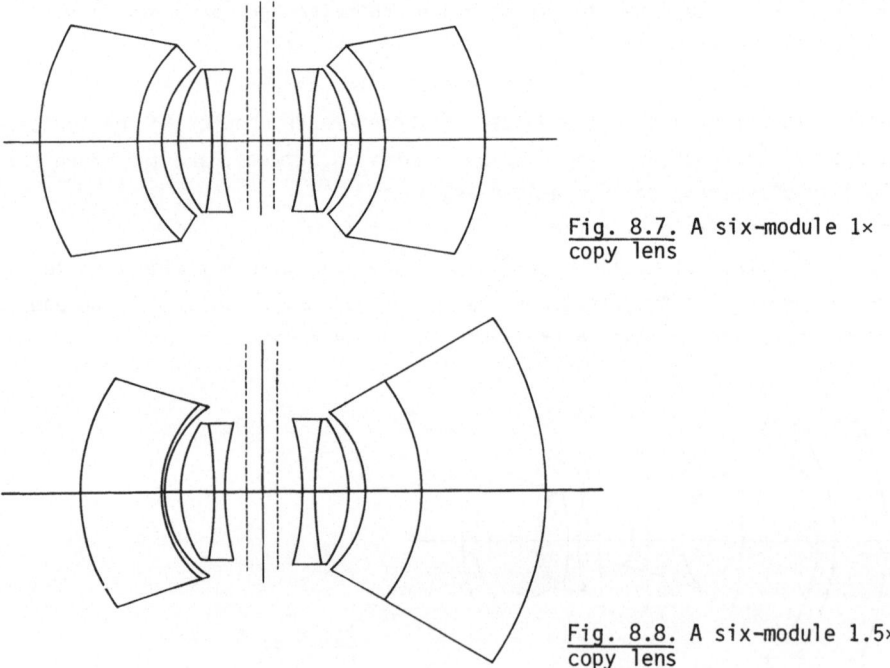

Fig. 8.7. A six-module 1×
copy lens

Fig. 8.8. A six-module 1.5×
copy lens

The design in Fig.8.7 and in Table 8.17 (see pp.177-178) is a six-module copy lens designed to operate at 1× with a 10 cm object height. It is symmetric and therefore possesses zero coma, distortion and lateral color. It also has been set to make axial color and the Petzval sum equal to zero.

The next design is obtained from this one by changing the magnification from 1× to 1.5× but keeping the same 10 cm object height. This is shown in Fig.8.8 and Table 8.18 (see pp.179-180). The lens-design equations were again used to bring all third-order aberrations back to zero.

8.4 Afocal Systems

Unlike the other two headings in this chapter this one is not a misnomer. Here we treat briefly the kind of lens that images infinity onto infinity; parallel rays are mapped into parallel rays. In a way, such systems can be said to have an infinite focal length. A synonym for *afocal* is *telescopic*.

Again, magnification is an important datum. Because focal length has no meaning, the idea of f number becomes somewhat obscure and we need to refer to an aperture diameter to fix the size of the opening. Semifield angle must also be specified.

An afocal system can be thought of as two camera lenses placed back to back, so that their image planes coincide. However, we must note that not both of the image planes need be real. We define magnification of the system as the ratio of the slope of the merging rays to the slope of the entering rays. Then magnification becomes the ratio of the focal lengths of the two camera lenses that constitute the system.

In terms of modules, an afocal system must consist of an even number of modules. The first must be in the backward orientation; the last, forward. The usual coupling relations must be observed. If we attempt to fix the magnification, we fix the ratio of the two focal lengths. This in turn fixes the ratio of the power parameters of the two "last" modules in the camera lenses.

But these two modules are hard-way coupled, and the ratio of the power parameters is already determined by the refractive indices and the two shape parameters by means of the coupling equation. We conclude that, in the case of the afocal lens, the magnification must be determined by means other than by setting the power parameters.

Suppose we take an eight-module afocal system as being composed of two camera lenses, specifically a three-module camera lens and a five-module camera lens. Referring to (6.8.16) and (6.8.14), the focal lengths of these two lenses would be given by

$$\mathscr{F}_A = f_3/M_A \qquad \mathscr{F}_B = f_4/M_B \quad , \tag{8.4.1}$$

where

$$M_A = E_1/E_2 \qquad M_B = E_8E_6/E_7E_5 \quad . \tag{8.4.2}$$

The magnification of the afocal system will be given by

$$M = \mathscr{F}_A/\mathscr{F}_B = (f_3/M_A)/(f_4/M_B) = (f_3/f_4)(E_8E_6E_2/E_7E_5E_1) \quad .$$

But, from (6.8.11),

$$f_3/f_4 = -E_4/E_3 \quad ,$$

so that

$$M = -E_8E_6E_4E_2/E_7E_5E_3E_1 \quad . \tag{8.4.3}$$

Thus, in such a system, we have no fixed scale of reference. If we applied the lens-design equations at this point they would be homogeneous. If they were solvable, we would obtain a system with four aberrations corrected. The aperture diameters would then be determined. It is perhaps more reasonable to set an entrance-pupil diameter and surrender one of the uncommitted power parameters. The lens-design equations are then nonhomogeneous and are more apt to lead to a usable solution.

Figure 8.9 shows a four-module afocal system. The same set of four modules is used in two designs. The first, shown in Table 8.19 (see pp.180-181) is corrected for zero coma, the second given in Table 8.20 (see pp.181-182) has zero axial color.

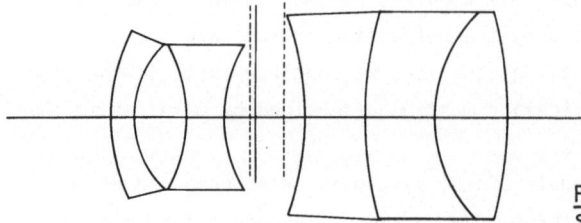

Fig. 8.9. A four-module afocal system

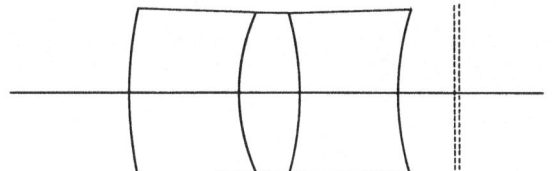

Fig. 8.10. A two-module Galilean telescope

Finally, we have a two-module Galilean telescope in Fig.8.10. The one uncommitted power parameter is used to fix the aperture diameter, so no aberrations are corrected except for third-order spherical and astigmatism. This design is shown in Table 8.21 (see p.183). These designs are from MERCADO [Ref.8.3, pp.102-115].

Additional examples of modular designs may be found in [8.4-7].

Table 8.1.

HYPOTHETIGON I					

Focal length: 10 cm   F/10
Semi-field angle: 30 degrees

SURFACE #1	T = INFINITY T = -3.026560E1 SPHERE	AIR C = 1.254020E-2 R # 7.974354E1		
SURFACE #2	T = 6.597830 SPHERE	INDEX C = -1.850580E-2 R = -5.403711E1	N = 1.921200	DEL N = 2.579000E-2
SURFACE #3	T = 2.051460E1 SPHERE	INDEX C = 6.195250E-2 R = 1.614140E1	N = 1.952500	DEL N = 4.677400E-2
SURFACE #4	T = 4.266540 SPHERE	INDEX C = 7.057500E-2 R = 1.416932E1	N = 1.464500	DEL N = 7.063000E-3
SURFACE #5	T = 3.952810 SPHERE	INDEX C = 1.231320E-1 R = 8.121366	N = 1.696800	DEL N = 1.257500E-2
SURFACE #6	T = 1.459890 SPHERE	INDEX C = 1.022190E-1 R = 9.782917	N = 1.464500	DEL N = 7.063000E-3
SURFACE #7	T = 2.538320 SPHERE	INDEX C = -2.578220E-2 R = -3.878645E1	N = 1.921200	DEL N = 2.579000E-2
SURFACE #8	T = 2.297270 SPHERE	INDEX C = 5.556410E-2 R = 1.799723E1	N = 1.531130	DEL N = 8.546000E-3
SURFACE #9	T = 1.511380 STOP	AIR		
SURFACE #10	T = 9.781090E-1 SPHERE	AIR C = -1.233710E-1. R = -8.171871		
SURFACE #11	T = 5.195620E-1 SPHERE	INDEX C = 5.947330E-2 R = 1.681427E1	N = 1.573100	DEL N = 1.345300E-2
SURFACE #12	T = 9.384140E-1 SPHERE	INDEX C = -2.656870E-1 R = -3.763827	N = 1.921200	DEL N = 2.579000E-2
SURFACE #13	T = 2.342920E-1 SPHERE	INDEX C = -3.249770E-1 R = -3.077141	N = 1.464500	DEL N = 7.063000E-3
SURFACE #14	T = 9.847700E-1 SPHERE	INDEX C = -2.415630E-1 R = -4.139707	N = 1.696800	DEL N = 1.257500E-2
SURFACE #15	T = 4.548530 SPHERE	INDEX C = -1.086170E-1 R = -9.206662	N = 1.952500	DEL N = 4.677400E-2
SURFACE #16	T = 2.021840E1 IMAGE	AIR		

							Marginal	Chief
Initial Rays						COORDS SLOPES	5.000000E-1 0.000000	0.000000 5.773503E-1
SURFACE #1 SPHAB COMA ASTIG	SPHERE 5.327983E-8 -3.043994E-6 1.739191E-4	PETZ CURVA DIST	-6.012925E-3 1.041801E-3 -5.952040E-2	LATCO LONGCO LAGRA	-4.208460E-5 2.404386E-3 -2.886751E-1	COORDS INCIDS SLOPES	5.000000E-1 6.270100E-3 3.006463E-3	1.747385E1 -3.582247E-1 4.055844E-1
SURFACE #2 SPHAB COMA ASTIG	SPHERE -5.327982E-8 -3.043988E-6 -1.739095E-4	PETZ CURVA DIST	1.544147E-4 -1.961973E-4 -1.120916E-2	LATCO LONGCO LAGRA	1.155421E-4 6.601162E-3 -2.886751E-1	COORDS INCIDS SLOPES	4.801639E-1 -1.189228E-2 2.815821E-3	1.479788E1 -6.794309E-1 3.946926E-1
SURFACE #3 SPHAB COMA ASTIG	SPHERE -7.351145E-6 -6.436029E-6 -5.634832E-6	PETZ CURVA DIST	1.057300E-2 -1.531716E-3 -1.341038E-3	LATCO LONGCO LAGRA	3.685013E-4 3.226280E-4 -2.886751E-1	COORDS INCIDS SLOPES	4.223984E-1 2.335282E-2 -4.965794E-3	6.700914 2.044571E-2 3.878797E-1

Table 8.1. (cont.)

SURFACE #4	SPHERE							
SPHAB	7.351124E-6	PETZ	-6.597510E-3	LATCO	-6.098654E-5	COORDS	4.435852E-1	5.046009
COMA	-6.436242E-6	CURVA	9.579038E-4	LONGCO	5.339648E-5	INCIDS	3.627182E-2	-3.175762E-2
ASTIG	5.635222E-6	DIST	-8.386882E-4	LAGRA	-2.886751E-1	SLOPES	-9.492794E-9	3.835320E-1

SURFACE #5	SPHERE							
SPHAB	-3.904036E-5	PETZ	1.151066E-2	LATCO	1.064031E-4	COORDS	4.435852E-1	3.529980
COMA	-3.654012E-5	CURVA	-1.695620E-3	LONGCO	9.958875E-5	INCIDS	5.461955E-2	5.112158E-2
ASTIG	-3.420001E-5	DIST	-1.587029E-3	LAGRA	-2.886751E-1	SLOPES	-8.663800E-3	3.754230E-1

SURFACE #6	SPHERE							
SPHAB	4.275173E-5	PETZ	-1.659207E-2	LATCO	-3.177983E-4	COORDS	4.562334E-1	2.981904
COMA	-5.459263E-5	CURVA	2.464573E-3	LONGCO	4.058186E-4	INCIDS	5.529952E-2	-7.061577E-2
ASTIG	6.971308E-5	DIST	-3.147183E-3	LAGRA	-2.886751E-1	SLOPES	4.481783E-3	3.586365E-1

SURFACE #7	SPHERE							
SPHAB	2.350592E-6	PETZ	-3.418833E-3	LATCO	-1.069142E-4	COORDS	4.448572E-1	2.071570
COMA	6.071979E-5	CURVA	2.061961E-3	LONGCO	-2.761775E-3	INCIDS	-1.595118E-2	-4.120461E-1
ASTIG	1.568495E-3	DIST	5.326397E-2	LAGRA	-2.886751E-1	SLOPES	8.545498E-3	4.636092E-1

SURFACE #8	SPHERE							
SPHAB	-1.981818E-6	PETZ	1.927450E-2	LATCO	5.480696E-5	COORDS	4.252259E-1	1.006534
COMA	5.357130E-5	CURVA	-4.230141E-3	LONGCO	-1.481509E-3	INCIDS	1.508180E-2	-4.076820E-1
ASTIG	-1.448107E-3	DIST	1.143466E-1	LAGRA	-2.886751E-1	SLOPES	5.351038E-4	6.801414E-1

SURFACE #9	STOP							
						COORDS	4.244172E-1	-2.141782E-2

SURFACE #10	SPHERE							
SPHAB	-2.486759E-5	PETZ	4.458129E-2	LATCO	1.899820E-4	COORDS	4.238938E-1	-6.866702E-1
COMA	-2.828587E-4	CURVA	-9.652156E-3	LONGCO	2.160968E-3	INCIDS	-5.240741E-2	-5.961129E-1
ASTIG	-3.217402E-3	DIST	-1.097893E-1	LAGRA	-2.886751E-1	SLOPES	-1.855757E-2	4.629700E-1

SURFACE #11	SPHERE							
SPHAB	2.308792E-5	PETZ	-6.850104E-3	LATCO	-1.473320E-4	COORDS	4.335356E-1	-9.272118E-1
COMA	-2.697748E-4	CURVA	4.140958E-3	LONGCO	1.721527E-3	INCIDS	4.434136E-2	-5.181144E-1
ASTIG	3.152231E-3	DIST	-4.838574E-2	LAGRA	-2.886751E-1	SLOPES	-1.052341E-2	3.690935E-1

SURFACE #12	SPHERE							
SPHAB	6.896549E-4	PETZ	-4.312602E-2	LATCO	-7.860897E-4	COORDS	4.434109E-1	-1.273574
COMA	1.974842E-4	CURVA	6.281255E-3	LONGCO	-2.250985E-4	INCIDS	-1.072851E-1	-3.072133E-2
ASTIG	5.655004E-5	DIST	1.798651E-3	LAGRA	-2.886751E-1	SLOPES	2.293313E-2	3.786738E-1

SURFACE #13	SPHERE							
SPHAB	-6.881505E-4	PETZ	3.037958E-3	LATCO	2.744313E-4	COORDS	4.380378E-1	-1.362295
COMA	2.666271E-4	CURVA	-4.488221E-3	LONGCO	-1.063297E-4	INCIDS	-1.652854E-1	6.404058E-2
ASTIG	-1.033059E-4	DIST	1.738982E-3	LAGRA	-2.886751E-1	SLOPES	3.047788E-4	3.874413E-1

SURFACE #14	SPHERE							
SPHAB	-1.751868E-4	PETZ	1.864402E-2	LATCO	1.303180E-3	COORDS	4.377377E-1	-1.743835
COMA	5.584509E-5	CURVA	-2.708834E-3	LONGCO	-4.154204E-4	INCIDS	-1.060460E-1	3.380476E-2
ASTIG	-1.780199E-5	DIST	8.635071E-4	LAGRA	-2.886751E-1	SLOPES	-1.358304E-2	3.918684E-1

SURFACE #15	SPHERE							
SPHAB	1.752371E-4	PETZ	-5.298729E-2	LATCO	-9.503164E-4	COORDS	4.995206E-1	-3.526260
COMA	3.815761E-5	CURVA	7.656366E-3	LONGCO	-2.069299E-4	INCIDS	-4.067339E-2	-8.856568E-3
ASTIG	8.308762E-6	DIST	1.667162E-3	LAGRA	-2.886751E-1	SLOPES	2.515836E-2	4.003042E-1

SURFACE #16	IMAGE							
						COORDS	-9.141231E-3	-1.161977E1

Sums of aberrations

SPHAB	3.855135E-6	PETZ	-4.673075E-4	LATCO	1.324601E-6
COMA	9.678613E-6	CURVA	1.019322E-4	LONGCO	8.572413E-3
ASTIG	3.448220E-5	DIST	-6.213967E-2	LAGRA	-2.886751E-1

Table 8.2.

Focal length: 20 cm   F/5
Semi field angle: degrees

	T = INFINITY			
	T = -2.452272E1	AIR		
SURFACE #1	SPHERE	C = 2.518358E-2		
		R = 3.970841E1		
	T = 5.353849	INDEX	N = 1.960520	DEL N = 2.750300E-2
SURFACE #2	SPHERE	C = 1.043960E-1		
		R = 9.578908		
	T = 7.133750	INDEX	N = 1.952500	DEL N = 4.677400E-2
SURFACE #3	SPHERE	C = 6.036440E-2		
		R = 1.656605E1		

Table 8.2. (cont.)

SURFACE #4	T = 4.623380 SPHERE	AIR C = 5.656537E-2 R = 1.767866E1			
SURFACE #5	T = 3.525231 SPHERE	INDEX C = 9.924384E-2 R = 1.007619E1	N = 1.464500	DEL N = 7.063000E-3	
SURFACE #6	T = 2.795093 SPHERE	AIR C = 7.141150E-2 R = 1.400335E1			
SURFACE #7	T = 4.002649 SPHERE	INDEX C =-2.651360E-2 R =-3.771649E1	N = 1.921200	DEL N = 2.579000E-2	
SURFACE #8	T = 2.317159 SPHERE	INDEX C = 5.348309E-2 R = 1.869750E1	N = 1.654730	DEL N = 1.991400E-2	
SURFACE #9	T = 2.800000 STOP	AIR			
SURFACE #10	T = 1.059790 SPHERE	AIR C =-7.988399E-2 R =-1.251815E1			
SURFACE #11	T = 1.601188 SPHERE	INDEX C = 3.706654E-2 R = 2.697851E1	N = 1.532560	DEL N = 1.158100E-2	
SURFACE #12	T = 2.127425 SPHERE	INDEX C =-1.453880E-1 R =-6.878146	N = 1.921200	DEL N = 2.579000E-2	
SURFACE #13	T = 4.281530E-1 SPHERE	INDEX C =-1.778323E-1 R =-5.623276	N = 1.464500	DEL N = 7.063000E-3	
SURFACE #14	T = 1.794120E-1 SPHERE	INDEX C =-1.983295E-1 R =-5.042115	N = 1.696800	DEL N = 1.240400E-2	
SURFACE #15	T = 3.034056 SPHERE	INDEX C =-9.585606E-2 R =-1.043231E1	N = 1.952500	DEL N = 4.677400E-2	
SURFACE #16	T = 2.257616E1 IMAGE	AIR			

Initial rays							Marginal	Chief
						COORDS	2.000000	0.000000
						SLOPES	0.000000	3.639702E-1
SURFACE #1	SPHERE							
SPHAB	4.386423E-5	PETZ	-1.233822E-2	LATCO	-1.413144E-3	COORDS	2.000000	8.925538
COMA	-1.212219E-4	CURVA	4.825751E-3	LONGCO	3.905322E-3	INCIDS	5.036716E-2	-1.391932E-1
ASTIG	3.350052E-4	DIST	-1.333630E-2	LAGRA	-7.279405E-1	SLOPES	2.467645E-2	2.957751E-1
SURFACE #2	SPHERE							
SPHAB	-4.386424E-5	PETZ	2.187238E-4	LATCO	-6.192092E-3	COORDS	1.867886	7.342003
COMA	-1.212219E-4	CURVA	-4.146142E-4	LONGCO	-1.711228E-2	INCIDS	1.703234E-1	4.707008E-1
ASTIG	-3.350053E-4	DIST	-1.145816E-3	LAGRA	-7.279405E-1	SLOPES	2.397683E-2	2.938417E-1
SURFACE #3	SPHERE							
SPHAB	-1.723629E-3	PETZ	2.944793E-2	LATCO	6.226581E-3	COORDS	1.696841	5.245810
COMA	-5.013330E-4	CURVA	-1.086399E-2	LONGCO	1.811057E-3	INCIDS	7.845198E-2	2.281846E-2
ASTIG	-1.458172E-4	DIST	-3.159889E-3	LAGRA	-7.279405E-1	SLOPES	-5.074867E-2	2.721071E-1
SURFACE #4	SPHERE							
SPHAB	1.723629E-3	PETZ	-1.794101E-2	LATCO	-1.490447E-3	COORDS	1.931472	3.987755
COMA	-5.013326E-4	CURVA	6.675812E-3	LONGCO	4.335095E-4	INCIDS	1.600031E-1	-4.653829E-2
ASTIG	1.458169E-4	DIST	-1.941718E-3	LAGRA	-7.279405E-1	SLOPES	0.000000	2.573464E-1
SURFACE #5	SPHERE							
SPHAB	-9.308993E-3	PETZ	3.147748E-2	LATCO	2.614987E-3	COORDS	1.931472	3.080549
COMA	-2.349465E-3	CURVA	-1.204984E-2	LONGCO	6.599874E-4	INCIDS	1.916867E-1	4.837914E-2
ASTIG	-5.929732E-4	DIST	-3.041217E-3	LAGRA	-7.279405E-1	SLOPES	-8.903846E-2	2.348743E-1
SURFACE #6	SPHERE							
SPHAB	9.308992E-3	PETZ	-3.434124E-2	LATCO	-7.163222E-3	COORDS	2.180342	2.424054
COMA	-2.349461E-3	CURVA	1.305576E-2	LONGCO	1.807898E-3	INCIDS	2.447400E-1	-6.176899E-2
ASTIG	5.929715E-4	DIST	-3.295095E-3	LAGRA	-7.279405E-1	SLOPES	2.831240	2.052566E-1

Table 8.2.(cont.)

```
SURFACE #7    SPHERE
SPHAB    3.787463E-4    PETZ   -2.222375E-3    LATCO   -4.585674E-4    COORDS    2.067018     1.602484
COMA     1.128924E-3    CURVA   4.173847E-3    LONGCO  -1.366846E-3    INCIDS   -8.311649E-2  -2.477442E-1
ASTIG    3.364968E-3    DIST    1.244093E-2    LAGRA   -7.279405E-1    SLOPES    4.169709E-2   2.451522E-1

SURFACE #8    SPHERE
SPHAB   -3.787464E-4    PETZ    2.116175E-2    LATCO    2.498943E-3    COORDS    1.970399     1.034427
COMA     1.128924E-3    CURVA  -1.106722E-2    LONGCO  -7.448562E-3    INCIDS    6.368594E-2  -1.898278E-1
ASTIG   -3.364967E-3    DIST    3.298789E-2    LAGRA   -7.279405E-1    SLOPES   -2.140837E-9   3.694381E-1

SURFACE #9    STOP
                                                                      COORDS    1.970399     6.547434E-7

SURFACE #10   SPHERE
SPHAB   -1.196751E-3    PETZ    2.775945E-2    LATCO    2.343671E-3    COORDS    1.970399    -3.915261E-1
COMA    -2.571070E-3    CURVA  -1.562724E-2    LONGCO   5.035085E-3    INCIDS   -1.574033E-1  -3.381614E-1
ASTIG   -5.523622E-3    DIST   -3.357316E-2    LAGRA   -7.279405E-1    SLOPES   -5.469719E-2   2.519280E-1

SURFACE #11   SPHERE
SPHAB    1.196751E-3    PETZ   -4.892597E-3    LATCO   -2.423802E-3    COORDS    2.057980    -7.949103E-1
COMA    -2.571068E-3    CURVA   7.304374E-3    LONGCO   5.2o7231E-3    INCIDS    1.309794E-1  -2.813926E-1
ASTIG    5.523614E-3    DIST   -1.569252E-2    LAGRA   -7.279405E-1    SLOPES   -2.820135E-2   1.950050E-1

SURFACE #12   SPHERE
SPHAB    2.310293E-2    PETZ   -2.359922E-2    LATCO   -9.789976E-3    COORDS    2.117976    -1.209769
COMA     1.579070E-3    CURVA   8.697342E-3    LONGCO  -6.691386E-4    INCIDS   -2.797269E-1  -1.911916E-2
ASTIG    1.079284E-4    DIST    5.944578E-4    LAGRA   -7.279405E-1    SLOPES    5.903067E-2   2.009673E-1

SURFACE #13   SPHERE
SPHAB   -2.310293E-2    PETZ    1.662416E-2    LATCO    3.287046E-3    COORDS    2.092702    -1.205814
COMA     1.579032E-3    CURVA  -6.158624E-3    LONGCO  -2.246621E-4    INCIDS   -4.311806E-1   2.947021E-1
ASTIG    1.o79233E-4    DIST    4.209279E-4    LAGRA   -7.279405E-1    SLOPES   -3.933621E-9   2.050019E-1

SURFACE #14   SPHERE
SPHAB   -1.984610E-2    PETZ    1.530722E-2    LATCO    2.453213E-2    COORDS    2.092702    -1.332593
COMA     2.835089E-3    CURVA  -5.976375E-3    LONGCO  -3.504506E-3    INCIDS   -4.150444E-1   5.929064E-2
ASTIG   -4.050030E-4    DIST    8.537474E-4    LAGRA   -7.279405E-1    SLOPES   -5.435435E-2   2.127666E-1

SURFACE #15   SPHERE
SPHAB    1.984610E-2    PETZ   -4.676205E-2    LATCO   -1.711230E-2    COORDS    2.257616    -1.978130
COMA     2.835124E-3    CURVA   1.742501E-2    LONGCO  -2.444586E-3    INCIDS   -1.620518E-1  -2.314999E-2
ASTIG    4.050131E-4    DIST    2.489258E-3    LAGRA   -7.279405E-1    SLOPES    1.000000E-2   2.348170E-1

SURFACE #16   IMAGE
                                                                      COORDS    1.499066E-7  -7.279405
```

Sums of Aberrations

```
SPHAB   -2.764864E-9    PETZ    6.839400E-10   LATCO   -4.540196E-3
COMA    -9.582436E-9    CURVA   5.929905E-9    LONGCO  -1.391049E-2
ASTIG    6.170126E-9    DIST   -2.539851E-2    LAGRA   -7.279405E-1
```

Table 8.3. Three modules used in the fifth-order study

Case	#1 Real	#2 Real	#3 Real
Shape parameter	0.162	-0.144	0.091
Orientation	backward	forward	backward
Category	AY	DY	AX
N_0	1.4645	1.4645	1.0
D_0	0.007063	0.007063	0.0
N_1	1.9212	1.0	1.0212
D_1	0.02579	0.0	0.02579
N_2	1.0	1.5168	1.5168
D_2	0.0	0.008054	0.008054

Table 8.4. ZSPHAB5

Focal length: 40 cm   F/10
Semi-field angle: 20 degrees

SURFACE #1	T = INFINITY T = -6.712709E1 SPHERE	AIR C = 1.588864E-2 R = 6.293806E1		
SURFACE #2	T = 3.243679E1 SPHERE	INDEX C = 2.897849E-2 R = 3.450836E1	N = 1.921200	DEL N = 2.579000E-2
SURFACE #3	T = 3.741946E1 SPHERE	INDEX C =-1.733317E-1 R =-5.769286	N = 1.464500	DEL N = 7.063000E-3
SURFACE #4	T = 6.078776E-1 SPHERE	AIR C =-1.851907E-1 R =-5.399838		
SURFACE #5	T = 3.497065E-1 SPHERE	INDEX C =-2.049784E-1 R =-4.878563	N = 1.501370	DEL N = 8.888000E-3
SURFACE #6	T = 9.856250E-1 SPHERE	INDEX C =-1.328114E-1 R =-7.529473	N = 1.921200	DEL N = 2.579000E-2
SURFACE #7	T = 2.504480E1 IMAGE	AIR		

Fifth Order Aberrations

Spherical	0.0	Petzval	-0.000162943
Coma	0.0148325	TOBSA	-0.00231810
Astigmatism	-0.00168904	SOBSA	0.000408691
Distortion	-0.00343217	ECOMA	0.00905326

Initial rays					Marginal	Chief
				COORDS	2.000000	0.000000
				SLOPES	0.000000	3.639702E-1

SURFACE #1 SPHERE						
SPHAB	1.100178E-5	PETZ	-7.618474E-3	LATCO	-8.531500E-4	COORDS	2.000000	2.443226E1
COMA	8.387114E-6	CURVA	2.779292E-3	LONGCO	-6.503919E-4	INCIDS	3.177727E-2	2.422514E-2
ASTIG	6.393848E-6	DIST	2.118770E-3	LAGRA	-7.279405E-1	SLOPES	1.523695E-2	3.755860E-1

SURFACE #2 SPHERE						
SPHAB	-1.100178E-5	PETZ	4.703756E-3	LATCO	7.065891E-4	COORDS	1.505762	1.224946E1
COMA	7.986681E-6	CURVA	-1.717825E-3	LONGCO	-5.129445E-4	INCIDS	2.839776E-2	-2.061520E-2
ASTIG	-5.797888E-6	DIST	1.247046E-3	LAGRA	-7.279405E-1	SLOPES	6.381189E-3	3.820148E-1

SURFACE #3 SPHERE						
SPHAB	1.019920E-2	PETZ	-5.497615E-2	LATCO	-2.022308E-3	COORDS	1.266982	-2.045326
COMA	1.240883E-3	CURVA	2.016065E-2	LONGCO	-2.460437E-4	INCIDS	-2.259893E-1	-2.749493E-2
ASTIG	1.509718E-4	DIST	2.452842E-3	LAGRA	-7.279405E-1	SLOPES	1.113532E-1	3.947862E-1

SURFACE #4 SPHERE						
SPHAB	-1.019920E-2	PETZ	6.184290E-2	LATCO	2.367410E-3	COORDS	1.199293	-2.285308
COMA	8.696332E-4	CURVA	-2.258312E-2	LONGCO	-2.018569E-4	INCIDS	-3.334510E-1	2.843165E-1
ASTIG	-7.414913E-5	DIST	1.925546E-3	LAGRA	-7.279405E-1	SLOPES	-2.190632E-9	4.042807E-1

SURFACE #5 SPHERE						
SPHAB	-3.137636E-3	PETZ	2.983467E-2	LATCO	3.321524E-3	COORDS	1.199293	-2.426688
COMA	1.188763E-3	CURVA	-1.130932E-2	LONGCO	-1.258433E-3	INCIDS	-2.458290E-1	9.313778E-2
ASTIG	-4.503892E-4	DIST	4.284787E-3	LAGRA	-7.279405E-1	SLOPES	-5.371976E-2	4.246336E-1

SURFACE #6 SPHERE						
SPHAB	3.137636E-3	PETZ	-6.368201E-2	LATCO	-3.636190E-3	COORDS	1.252240	-2.845217
COMA	1.302973E-3	CURVA	2.372944E-2	LONGCO	-1.510008E-3	INCIDS	-1.125920E-1	-4.675632E-2
ASTIG	5.410883E-4	DIST	9.850023E-3	LAGRA	-7.279405E-1	SLOPES	5.000000E-2	4.677055E-1

SURFACE #7   IMAGE

Sums of Aberrations

SPHAB	-4.765752E-10	PETZ	-2.989531E-2	LATCO	-1.161240E-4
COMA	4.618626E-3	CURVA	1.104912E-2	LONGCO	-4.379678E-3
ASTIG	1.681177E-4	DIST	2.187901E-2	LAGRA	-7.279405E-1

Table 8.5. ZTOBSA

Focal length:40 cm   F/10
Semi-field angle:20 degrees

SURFACE #1	T = INFINITY T = -5.048265E1 SPHERE	AIR C = 2.109574E-2 R = 4.740294E1					
SURFACE #2	T = 2.443035 SPHERE	INDEX C = 3.847547E-2 R = 2.599059E1	N = 1.921200	DEL N = 2.579000E-2			
SURFACE #3	T = 2.818313E1 SPHERE	INDEX C =-2.301369E-1 R =-4.345240	N = 1.464500	DEL N = 7.063000E-3			
SURFACE #4	T = 4.578338E-1 SPHERE	AIR C =-2.458824E-1 R =-4.066985					
SURFACE #5	T = 1.758320 SPHERE	INDEX C =-2.049784E-1 R =-4.878563	N = 1.501370	DEL N = 8.888000E-3			
SURFACE #6	T = 9.856250E-1 SPHERE	INDEX C =-1.328114E-1 R =-7.529473	N = 1.921200	DEL N = 2.579000E-2			
SURFACE #7	T = 2.504480E1 IMAGE	AIR					

Fifth-Order Aberrations

Spherical	-0.00296154	Petzval	-0.000281714
Coma	0.0342160	TOBSA	0.0
Astigmatism	-0.00161526	SOBSA	0.00394767
Distortion	-0.00346147	ECOMA	0.00982271

							Marginal	Chief
Initial rays						COORDS SLOPES	2.000000 0.000000	0.000000 3.639702E-1
SURFACE #1 SPHERE								
SPHAB	2.575059E-5	PETZ	-1.011524E-2	LATCO	-1.132749E-3	COORDS	2.000000	1.837418E1
COMA	1.443225E-5	CURVA	3.689734E-3	LONGCO	-6.348634E-4	INCIDS	4.219148E-2	2.364675E-2
ASTIG	8.088737E-6	DIST	2.067959E-3	LAGRA	-7.279405E-1	SLOPES	2.023048E-2	3.753087E-1
SURFACE #2 SPHERE								
SPHAB	-2.575064E-5	PETZ	6.245295E-3	LATCO	9.381564E-4	COORDS	1.505762	9.205261
COMA	1.443226E-5	CURVA	-2.281190E-3	LONGCO	-5.258014E-4	INCIDS	3.770444E-2	-2.113192E-2
ASTIG	-8.088740E-6	DIST	1.278521E-3	LAGRA	-7.279405E-1	SLOPES	8.472459E-3	3.818986E-1
SURFACE #3 SPHERE								
SPHAB	2.387213E-2	PETZ	-7.299322E-2	LATCO	-2.685071E-3	COORDS	1.266982	-1.557837
COMA	1.860350E-3	CURVA	2.671234E-2	LONGCO	-2.092470E-4	INCIDS	-3.000517E-1	-2.338297E-2
ASTIG	1.449767E-4	DIST	2.081687E-3	LAGRA	-7.279405E-1	SLOPES	1.478465E-1	3.927600E-1
SURFACE #4 SPHERE								
SPHAB	-2.387213E-2	PETZ	8.211038E-2	LATCO	3.143272E-3	COORDS	1.199293	-1.737655
COMA	1.860185E-3	CURVA	-3.003069E-2	LONGCO	-2.449328E-4	INCIDS	-4.427315E-1	3.449891E-2
ASTIG	-1.449510E-4	DIST	2.340077E-3	LAGRA	-7.279405E-1	SLOPES	-1.332774E-8	4.042806E-1
SURFACE #5 SPHERE								
SPHAB	-3.137639E-3	PETZ	2.983467E-2	LATCO	3.321526E-3	COORDS	1.199293	-2.448510
COMA	1.245858E-3	CURVA	-1.135362E-2	LONGCO	-1.318873E-3	INCIDS	-2.458291E-1	9.761098E-2
ASTIG	-4.946908E-4	DIST	4.508165E-3	LAGRA	-7.279405E-1	SLOPES	-5.371979E-2	4.256110E-1
SURFACE #6 SPHERE								
SPHAB	3.137639E-3	PETZ	-6.368201E-2	LATCO	-3.636191E-3	COORDS	1.252240	-2.868003
COMA	1.245879E-3	CURVA	2.367306E-2	LONGCO	-1.443842E-3	INCIDS	-1.125920E-1	-4.470753E-2
ASTIG	4.947081E-4	DIST	9.399992E-3	LAGRA	-7.279405E-1	SLOPES	5.000000E-2	4.667956E-1
SURFACE #7 IMAGE								
						COORDS	5.114289E-7	-1.455881E1

Sum of Aberrations

SPHAB	-1.903572E-9	PETZ	-2.860012E-2	LATCO	-5.105625E-5
COMA	6.241137E-3	CURVA	1.040964E-2	LONGCO	-4.377560E-3
ASTIG	4.306116E-8	DIST	2.167640E-2	LAGRA	-7.279405E-1

Table 8.6. ZASTIG5

| Focal length:40 cm   F/10 | | | | | | | |
| Semi-field angle:20 degrees | | | | | | | |

SURFACE #1	T = INFINITY T = -7.402105 SPHERE		AIR C = 1.438738E-1 R = 6.950534				
SURFACE #2	T = 3.582142 SPHERE		INDEX C = 2.624043E-1 R = 3.810914	N = 1.921200	DEL N = 2.579000E-2		
SURFACE #3	T = 4.132400 SPHERE		INDEX C =-1.569543 R =-6.371283E-1	N = 1.464500	DEL N = 7.063000E-3		
SURFACE #4	T = 6.713066E-2 SPHERE		AIR C =-1.677181 R =-5.962385E-1				
SURFACE #5	T = 5.426248 SPHERE		INDEX C =-2.049784E-1 R =-4.878563	N = 1.501370	DEL N = 8.888000E-3		
SURFACE #6	T = 9.856250E-1 SPHERE		INDEX C =-1.328114E-1 R =-7.529473	N = 1.921200	DEL N = 2.579000E-2		
SURFACE #7	T = 2.504480E1 IMAGE		AIR				

Fifth-Order Aberrations

Spherical	-57.6744	Petzval	-0.00320656
Coma	61.2932	TOBSA	1.95719
Astigmatism	0.0	SOBSA	2.07885
Distortion	-0.00353775	ECOMA	0.0976487

							Marginal	Chief
Initial rays						COORDS SLOPES	2.000000 0.000000	0.000000 3.639702E-1
SURFACE #1	SPHERE							
SPHAB	8.168613E-3	PETZ	-6.898635E-2	LATCO	-7.725393E-3	COORDS	2.000000	2.694146
COMA	6.712893E-4	CURVA	2.516414E-2	LONGCO	-6.348659E-4	INCIDS	2.877477E-1	2.364685E-2
ASTIG	5.516596E-5	DIST	2.067967E-3	LAGRA	-7.279405E-1	SLOPES	1.379727E-1	3.753087E-1
SURFACE #2	SPHERE							
SPHAB	-8.168612E-3	PETZ	4.259317E-2	LATCO	6.398262E-3	COORDS	1.505762	1.349737
COMA	6.712895E-4	CURVA	-1.555781E-2	LONGCO	-5.258037E-4	INCIDS	2.571457E-1	-2.113201E-2
ASTIG	-5.516599E-5	DIST	1.278527E-3	LAGRA	-7.279405E-1	SLOPES	5.778256E-2	3.818987E-1
SURFACE #3	SPHERE							
SPHAB	7.572715	PETZ	-4.978167E-1	LATCO	-1.831228E-2	COORDS	1.266982	-2.284210E-1
COMA	8.652709E-2	CURVA	1.821791E-1	LONGCO	-2.092392E-4	INCIDS	-2.046364	-2.338210E-2
ASTIG	9.886728E-4	DIST	2.081609E-3	LAGRA	-7.279405E-1	SLOPES	1.008319	3.927596E-1
SURFACE #4	SPHERE							
SPHAB	-7.574745	PETZ	5.600806E-1	LATCO	2.143939E-2	COORDS	1.199293	-2.547873E-1
COMA	8.670219E-2	CURVA	-2.048451E-1	LONGCO	-2.453999E-4	INCIDS	-3.019749	3.456471E-2
ASTIG	-9.924122E-4	DIST	2.344702E-3	LAGRA	-7.279405E-1	SLOPES	-1.013988E-4	4.043022E-1
SURFACE #5	SPHERE							
SPHAB	-3.137855E-3	PETZ	2.983467E-2	LATCO	3.323202E-3	COORDS	1.199843	-2.448631
COMA	1.245928E-3	CURVA	-1.135364E-2	LONGCO	-1.319522E-3	INCIDS	-2.458404E-1	9.761426E-2
ASTIG	-4.947123E-4	DIST	4.508117E-3	LAGRA	-7.279405E-1	SLOPES	-5.382365E-2	4.256334E-1
SURFACE #6	SPHERE							
SPHAB	3.135668E-3	PETZ	-6.368201E-2	LATCO	-3.637528E-3	COORDS	1.252893	-2.868146
COMA	1.245379E-3	CURVA	2.367298E-2	LONGCO	-1.444701E-3	INCIDS	-1.125748E-1	-4.471082E-2
ASTIG	4.946215E-4	DIST	9.402089E-3	LAGRA	-7.279405E-1	SLOPES	4.988026E-2	4.668210E-1
SURFACE #7	IMAGE							
						COORDS	3.651567E-3	-1.455959E1

Sum of Aberrations

SPHAB	-2.032650E-3	PETZ	2.023411E-3	LATCO	1.485648E-3
COMA	1.770632E-1	CURVA	-7.402915E-4	LONGCO	-4.379531E-3
ASTIG	-3.830227E-6	DIST	2.168301E-2	LAGRA	-7.279405E-1

Table 8.7. Three module camera-zero TOBSA

Focal length:40 mm   F/10
Semi-field angle:30 degrees

	T = INFINITY				
	T =-3.464444E1	AIR			
SURFACE #1	SPHERE	C = 2.892149E-2			
		R = 3.457636E1			
	T = 1.419169E1	INDEX	N = 1.921200	DEL N = 2.579000E-2	
SURFACE #2	SPHERE	C = 4.888683E-2			
		R = 2.045541E1			
	T = 2.350676E1	INDEX	N = 1.464500	DEL N = 7.063000E-3	
SURFACE #3	SPHERE	C =-3.226500E-1			
		R =-3.099333			
	T = 5.015393E-2	AIR			
SURFACE #4	SPHERE	C =-3.180026E-1			
		R =-3.144629			
	T = 1.934812	INDEX	N = 1.516800	DEL N = 8.054000E-3	
SURFACE #5	SPHERE	C =-1.961720E-1			
		R =-5.097567			
	T = 4.113833	INDEX	N = 1.921200	DEL N = 2.579000E-2	
SURFACE #6	SPHERE	C =-1.088700E-1			
		R =-9.185271			
	T = 3.078961E1	AIR			
SURFACE #7	IMAGE				

Fifth Order Aberrations		Third Order Pupil Aberrations	
Spherical	-0.0125606	Spherical	0.0
Coma	0.00101978	Coma	0.0
Astigmatism	0.0	Astigmatism	-0.0181985
Distortion	0.0	Distortion	0.00242876
Petzval	0.00436951		
TOBSA	0.0		
SOBSA	0.00001867		
ECOMA	0.00230331		

							Marginal	Chief
Initial rays								
						COORDS	2.000000	0.000000
						SLOPES	0.000000	5.773503E-1

SURFACE #1	SPHERE							
SPHAB	4.183035E-5	PETZ	-1.386762E-2	LATCO	-1.552957E-3	COORDS	2.000000	2.000198E1
COMA	8.220462E-7	CURVA	8.006492E-3	LONGCO	-3.051857E-5	INCIDS	5.784298E-2	1.136725E-3
ASTIG	1.615478E-8	DIST	1.573428E-4	LAGRA	-1.154701	SLOPES	2.773525E-2	5.778953E-1
SURFACE #2	SPHERE							
SPHAB	-4.183034E-5	PETZ	7.935255E-3	LATCO	1.348365E-3	COORDS	1.606390	1.180067E1
COMA	8.220325E-7	CURVA	-4.521438E-3	LONGCO	-2.649751E-5	INCIDS	5.079606E-2	-9.982232E-4
ASTIG	-1.615424E-8	DIST	9.003236E-5	LAGRA	-1.154701	SLOPES	1.189464E-2	5.782066E-1
SURFACE #3	SPHERE							
SPHAB	4.964946E-2	PETZ	-1.023359E-1	LATCO	-4.123109E-3	COORDS	1.326785	-1.791096
COMA	3.493013E-5	CURVA	5.908369E-2	LONGCO	-2.900752E-6	INCIDS	-4.399820E-1	-3.095428E-4
ASTIG	2.457457E-8	DIST	4.156749E-5	LAGRA	-1.154701	SLOPES	2.162663E-1	5.783504E-1
SURFACE #4	SPHERE							
SPHAB	-4.964946E-2	PETZ	1.083490E-1	LATCO	4.435205E-3	COORDS	1.315939	-1.820102
COMA	3.494601E-5	CURVA	-6.255533E-2	LONGCO	-3.121740E-6	INCIDS	-6.347382E-1	4.467636E-4
ASTIG	-2.459692E-8	DIST	4.402985E-5	LAGRA	-1.154701	SLOPES	0.000000	5.785026E-1
SURFACE #5	SPHERE							
SPHAB	-2.471027E-3	PETZ	2.722371E-2	LATCO	4.180940E-3	COORDS	1.315939	-2.939396
COMA	-1.795045E-5	CURVA	-1.571775E-2	LONGCO	3.037190E-5	INCIDS	-2.581504E-1	-1.875300E-3
ASTIG	-1.303988E-7	DIST	-1.141795E-4	LAGRA	-1.154701	SLOPES	-5.433897E-2	5.781079E-1
SURFACE #6	SPHERE							
SPHAB	2.471027E-3	PETZ	-5.220227E-2	LATCO	-4.496950E-3	COORDS	1.539480	-5.317635
COMA	-1.795139E-5	CURVA	3.013913E-2	LONGCO	3.266922E-5	INCIDS	-1.132642E-1	8.228361E-4
ASTIG	1.304124E-7	DIST	-2.189535E-4	LAGRA	-1.154701	SLOPES	5.000000E-2	5.773499E-1
SURFACE #7	IMAGE							
						COORDS	-2.397610E-8	-2.309401E1

Sum of Aberrations								
SPHAB	8.813004E-10	PETZ	-2.489786E-2	LATCO	-2.085071E-4			
COMA	3.561838E-5	CURVA	1.437479E-2	LONGCO	2.554955E-9			
ASTIG	-8.200385E-12	DIST	-1.604417E-7	LAGRA	-1.154701			

Table 8.8. ZCOPE13

Focal length:40 mm   F/4.02002159
Semi field angle: 20 degrees

SURFACE #1	T = INFINITY T =-3.057818E1 SPHERE	AIR C = 2.473235E-2 R = 4.043288E1		CLEAR AP = 1.504754E1
SURFACE #2	T = 4.978174 SPHERE	INDEX C = 4.647410E-1 R = 2.151736E1	N = 1.808020	DEL N = 1.984100E-2 CLEAR AP = 1.268509E1
SURFACE #3	T = 1.717455 SPHERE	INDEX C =-2.610237E-2 R =-3.831070E1	N = 1.670030	DEL N = 1.422200E-2 CLEAR AP = 1.456307E1
SURFACE #4	T = 2.099632E1 SPHERE	INDEX C = 3.546460E-2 R = 2.819713E1	N = 1.603110	DEL N = 9.952000E-3 CLEAR AP = 4.888877
SURFACE #5	T = 3.772200 STOP	AIR		CLEAR AP = 3.034943
SURFACE #6	T = 1.323622 SPHERE	AIR C =-1.075259E-1 R =-9.300085		CLEAR AP = 3.270508
SURFACE #7	T = 6.135148E-1 SPHERE	INDEX C = 7.154052E-2 R = 1.397809E1	N = 1.464500	DEL N = 7.063000E-3 CLEAR AP = 4.174378
SURFACE #8	T = 1.748647 SPHERE	INDEX C =-2.081393E-1 R =-4.804474	N = 1.668820	DEL N = 1.165500E-2 CLEAR AP = 3.870135
SURFACE #9	T = 1.751234E-1 SPHERE	AIR C =-2.327267E-1 R =-4.296886		CLEAR AP = 4.132767
SURFACE #10	T = 7.557396E-1 SPHERE	INDEX C =-1.872254E-1 R =-5.341156	N = 1.464500	DEL N = 7.063000E-3 CLEAR AP = 4.736842
SURFACE #11	T = 3.754405 SPHERE	INDEX C =-1.111730E-1 R =-8.994986	N = 1.952500	DEL N = 4.677400E-2 CLEAR AP = 7.212567
SURFACE #12	T =-9.094008 PUPIL	AIR		
SURFACE #13	T = 3.548697E1 IMAGE	AIR		

			Marginal	Chief
Initial rays				
		COORDS	4.975098	0.000000
		SLOPES	0.000000	3.639702E-1

SURFACE #1 SPHERE							
SPHAB	6.325875E-4	PETZ	-1.105310E-1	LATCO	-6.717828E-3	COORDS	4.975098	1.112955E1
COMA	-4.560665E-4	CURVA	1.033621E-2	LONGCO	4.843245E-3	INCIDS	1.230458E-1	-8.871039E-2
ASTIG	3.288029E-4	DIST	-7.451934E-3	LAGRA	-1.810787	SLOPES	5.499026E-2	3.243248E-1

SURFACE #2 SPHERE							
SPHAB	-6.325875E-4	PETZ	2.123885E-3	LATCO	3.415886E-3	COORDS	4.701347	9.515003
COMA	-4.560665E-4	CURVA	-2.251755E-3	LONGCO	2.462697E-3	INCIDS	1.635006E-1	1.178764E-1
ASTIG	-3.288029E-4	DIST	-1.623412E-3	LAGRA	-1.810787	SLOPES	4.148066E-2	3.145850E-1

SURFACE #3 SPHERE							
SPHAB	4.946462E-4	PETZ	-6.524515E-4	LATCO	-2.897242E-3	COORDS	4.630106	8.974720
COMA	1.672349E-3	CURVA	6.244771E-3	LONGCO	-9.795286E-3	INCIDS	-1.623374E-1	-5.488464E-1
ASTIG	5.654046E-3	DIST	2.111295E-2	LAGRA	-1.810787	SLOPES	4.825725E-2	3.374959E-1

SURFACE #4 SPHERE							
SPHAB	-4.946462E-4	PETZ	1.334223E-2	LATCO	2.880120E-3	COORDS	3.616881	1.888548
COMA	1.672349E-3	CURVA	-1.773401E-2	LONGCO	-9.737399E-3	INCIDS	8.001401E-2	-2.705194E-1
ASTIG	-5.654946E-3	DIST	5.995693E-2	LAGRA	-1.810787	SLOPES	1.054218E-9	5.006489E-1

SURFACE #5 STOP				
			COORDS	3.616881	1.072840E-8

SURFACE #6 SPHERE							
SPHAB	-1.272289E-2	PETZ	3.410432E-2	LATCO	6.783930E-3	COORDS	3.616881	-6.626697E-1
COMA	-1.404737E-2	CURVA	-4.638758E-2	LONGCO	7.490154E-3	INCIDS	-3.889084E-1	-4.293947E-1
ASTIG	-1.550974E-2	DIST	-5.121664E-2	LAGRA	-1.810787	SLOPES	-1.233513E-1	3.644564E-1

Table 8.8. (cont.)

SURFACE #7	SPHERE							
SPHAB	1.272289E-2	PETZ	-5.980867E-3	LATCO	-4.528962E-3	COORDS	3.692559	-8.862681E-
COMA	-1.404737E-2	CURVA	2.092478E-2	LONGCO	5.000438E-3	INCIDS	3.875189E-1	-4.278606E-
ASTIG	1.550974E-2	DIST	-2.310310E-2	LAGRA	-1.810787	SLOPES	-7.590586E-2	3.120718E-
SURFACE #8	SPHERE							
SPHAB	6.887859E-1	PETZ	-8.341687E-2	LATCO	-3.211314E-2	COORDS	3.825291	-1.431973
COMA	1.340877E-2	CURVA	7.578614E-2	LONGCO	-6.251546E-4	INCIDS	-7.202877E-1	-1.402202E-
ASTIG	2.610318E-4	DIST	1.475348E-3	LAGRA	-1.810787	SLOPES	4.058370E-1	3.214500E-
SURFACE #9	SPHERE							
SPHAB	-6.887859E-1	PETZ	7.381463E-2	LATCO	2.316727E-2	COORDS	3.754220	-1.488266
COMA	1.340873E-2	CURVA	-6.709234E-2	LONGCO	-4.510018E-4	INCIDS	-1.279544	2.490914E-
ASTIG	-2.610306E-4	DIST	1.306100E-3	LAGRA	-1.810787	SLOPES	2.423712E-8	3.293505E-
SURFACE #10	SPHERE							
SPHAB	-9.883073E-2	PETZ	3.195245E-2	LATCO	7.394008E-2	COORDS	3.754220	-1.737169
COMA	-5.776677E-4	CURVA	-2.893292E-2	LONGCO	4.321814E-4	INCIDS	-7.028852E-1	-4.108379E-.
ASTIG	-3.376480E-6	DIST	-1.691136E-4	LAGRA	-1.810787	SLOPES	-1.756763E-1	3.283237E-
SURFACE #11	SPHERE							
SPHAB	9.883074E-2	PETZ	-5.423423E-2	LATCO	-6.503532E-2	COORDS	4.413780	-2.969829
COMA	-5.776650E-4	CURVA	4.910671E-2	LONGCO	3.801310E-4	INCIDS	-3.150171E-1	1.841273E-:
ASTIG	3.376448E-6	DIST	-2.870284E-4	LAGRA	-1.810787	SLOPES	1.243775E-1	3.265699E-:
SURFACE #12	PUPIL							
						COORDS	5.544870	1.586886E-£
SURFACE #13	IMAGE							
						COORDS	1.131090	-1.158898E1

Sums of Aberrations

SPHAB	1.120497E-8	PETZ	-3.448804E-9	LATCO	-1.105207E-3
COMA	-6.536084E-9	CURVA	5.544280E-9	LONGCO	5.352035E-9
ASTIG	2.436730E-9	DIST	9.427015E-8	LAGRA	-1.810787

TABLE 8.9   FIRST MODULE

MODULE PRINT OUT

N(0) = 1.67003 D(0) = .014222
N(1) = 1.4645   D(1) = 7.063E-03
N(2) = 1   D(2) = 0

CRITICAL VALUES

REAL CASE		IMAGINARY CASE	
-1	MINUS ONE	0	ZERO
-.300251665	PSI 0	60	PHI 0
-.116418557	PSI-BAR-PRIME	180	PI
-.108637671	PSI*		
-.10448466	PSI 0-TILDE		
-.0869407656	PSI-C		
-.0740775145	PSI-P		
-.0447794706	PSI INF		
0	ZERO		
.0856377604	PSI-P		
1	PLUS ONE		

MODULE: TABLE 8.9 FIRST MODULE
FILE: ONE
REAL CASE. PSI = -.109 CATEGORY: DX

Q	2.391010	SPHERICAL AB	0.000000
T0	1.686752	COMA	1.747179
C1	2.079680	ASTIGMATISM	3.497007E-10
T1	-1.024226E-2	PETZVAL	-8.028481E-1
C2	-3.082275	DISTORTION	2.913468E-1
E	1.596664	LATERAL COL	-8.727887E-3
G	-7.722047E-2	LONG COLOR	1.950119E-3
A	-4.893899E-2	CONCENTRIC	-8.155212E-1

TABLE 8.10 SECOND MODULE

MODULE PRINT OUT

N(0) = 1.67003 D(0) = .014222
N(1) = 1 D(1) = 0
N(2) = 1.4645 D(2) = 7.063E-03

CRITICAL VALUES

REAL CASE		IMAGINARY CASE	
-1	MINUS ONE	0	ZERO
-.420983197	PSI 0	60	PHI 0
-.207981233	PSI 0-TILDE	180	PI
-.189096137	PSI INF		
-.166735466	PSI-P		
-.161397643	PSI*		
-.160174441	PSI-C		
-.132476011	PSI-BAR-PRIME		
0	ZERO		
.216788542	PSI-P		
1	PLUS ONE		

MODULE: TABLE 8.10 SECOND MODULE
FILE: TWO
REAL CASE. PSI = -.162 CATEGORY: AY

Q	-5.221909	SPHERICAL AB	-1.287553E-7
T0	1.688639	COMA	1.305325
C1	-6.091305	ASTIGMATISM	-4.455387E-10
T1	3.563681E-3	PETZVAL	-3.087949E-1
C2	-6.731616	DISTORTION	9.858168E-3
E	1.530472	LATERAL COL	-5.476818E-4
G	-1.543863E-1	LONG COLOR	-3.241564E-2
A	-1.548432E-1	CONCENTRIC	1.917938E-2

TABLE 8.11   THIRD MODULE

MODULE PRINT OUT

N(0) = 1   D(0) = 0
N(1) = 1.9525 D(1) = .046774
N(2) = 1.4645 D(2) = 7.063E-03

CRITICAL VALUES

REAL CASE		IMAGINARY CASE	
-1	MINUS ONE	0	ZERO
-.149066088	PSI 0	60	PHI 0
-.0853156917	PSI-P	180	PI
0	ZERO		
.0766641085	PSI 0-TILDE		
.100839262	PSI*		
.106169135	PSI-C		
.1234792	PSI-P		
.137520797	PSI-BAR-PRIME		
.202122965	PSI INF		
1	PLUS ONE		

MODULE: TABLE 8.11   THIRD MODULE
  FILE: THREE
REAL CASE.   PSI = .102   CATEGORY: AX

Q	-2.524456	SPHERICAL AB	0.000000
T0	1.035006	COMA	-1.143359
C1	3.575068	ASTIGMATISM	1.649880E-10
T1	2.707458E-2	PETZVAL	-8.611965E-1
C2	5.173065	DISTORTION	-4.054734E-2
E	1.264701	LATERAL COL	7.866474E-3
G	2.516480E-1	LONG COLOR	2.841210E-2
A	2.373657E-1	CONCENTRIC	-5.933136E-2

Table 8.12. Printout from Program 'Camera.4'

FILE: ONE	FILE: TWO	FILE: THREE
PSI = -.109	PSI = -.162	PSI = .102

FILE: ONE	FILE: TWO	FILE: THREE
Q = 2.39100962	Q = -5.22190926	Q = -2.52445552
T0 = 1.68675199	T0 = 1.68863922	T0 = 1.03500568
C1 = 2.07968008	C1 = -6.09130465	C1 = 3.57506809
T1 = -.0102422631	T1 = 3.56368135E-03	T1 = .0270745846
C2 = -3.08227463	C2 = -6.73161594	C2 = 5.17306458
E = 1.59666388	E = 1.53047168	E = 1.26470063
G = -.0772204744	G = -.154386343	G = .251648034
A = -.0489389948	A = -.154843247	A = .2373657
CATEGORY: DX	CATEGORY: AY	CATEGORY: AX

FILE: ONE	FILE: TWO	FILE: THREE
DISTORTION = .291346781	DISTORTION = 9.85816775E-03	DISTORTION = -.0405473395
COMA = 1.74717902	COMA = 1.30532488	COMA = -1.14335912
AXIAL COL = 1.95011878E-03	AXIAL COL = -.032415635	AXIAL COL = .0284121009
LATRL COL = -8.72788693E-03	LATRL COL = -5.47681762E-04	LATRL COL = 7.86647447E-03
PETZVAL = -.802848122	PETZVAL = -.308794887	PETZVAL = -.861196522
CONCENTRIC = -.815521175	CONCENTRIC = .0191793831	CONCENTRIC = -.0593313565

PSI CHANGE TO SET LATERAL COLOR EQUAL TO ZERO

PSI = -.109159	PSI = -.161674	PSI = .101765

Q = 2.38860075	Q = -5.14911932	Q = -2.51261982
T0 = 1.69412503	T0 = 1.67855097	T0 = 1.02785337
C1 = 2.06371088	C1 = -6.06319336	C1 = 3.58785596
T1 = -.0147731546	T1 = 1.65484074E-03	T1 = .0216442219
C2 = -3.07916934	C2 = -6.63778169	C2 = 5.1488111
E = 1.60040132	E = 1.51856736	E = 1.25402856
G = -.0802782212	G = -.156832356	G = .249081992
A = -.0477856716	A = -.155915513	A = .241183265
CATEGORY: DX	CATEGORY: AY	CATEGORY: AX

DISTORTION = .293620115	DISTORTION = 8.16718906E-03	DISTORTION = -.0433892393
COMA = 1.7512646	COMA = 1.0470288	COMA = -1.20173864
AXIAL COL = 1.99468741E-03	AXIAL COL = -.0316708107	AXIAL COL = .0288438094
LATRL COL = -8.77883869E-03	LATRL COL = -4.50762571E-04	LATRL COL = 8.32809572E-03
PETZVAL = -.803205183	PETZVAL = -.327278099	PETZVAL = -.871574082
CONCENTRIC = -.824100068	CONCENTRIC = .0159316917	CONCENTRIC = -.0628541738

Table 8.12.(cont.)

GENERATED DESIGNS. ZPETZ

T = .854001204
AIR N = 1 DEL N = 0
SPHERE C = -.172295044
T = .2640183
INDEX N = 1.4645 DEL N = 7.063E-03
SPHERE C = .115475025
T = 1.33823965
INDEX N = 1.67003 DEL N = .014222
SPHERE C = -.321918391
T = .0311682081
AIR N = 1 DEL N = 0
SPHERE C = -.352425509
T = 4.68881004
INDEX N = 1.4645 DEL N = 7.063E-03
SPHERE C = -.162851163
T = .684318171
INDEX N = 1.9525 DEL N = .046774
SPHERE C = -.113479889
T = 32.4972984
AIR N = 1 DEL N = 0
IMAGE C = 0
F1 = -17.8714911 F2 = 31.6166678

ZCOMA

T = 1.09742029
AIR N = 1 DEL N = 0
SPHERE C = -.134078234
T = .339272402
INDEX N = 1.4645 DEL N = 7.063E-03
SPHERE C = .0898614788
T = 1.71968298
INDEX N = 1.67003 DEL N = .014222
SPHERE C = -.250513587
T = .0400521965
AIR N = 1 DEL N = 0
SPHERE C = -.274253913
T = 3.85177988
INDEX N = 1.4645 DEL N = 7.063E-03
SPHERE C = -.162851163
T = .684318171
INDEX N = 1.9525 DEL N = .046774
SPHERE C = -.113479889
T = 32.4972984
AIR N = 1 DEL N = 0
IMAGE C = 0
F1 = -22.9654676 F2 = 31.6166678

ZAXIAL

T = 1.66814429
AIR N = 1 DEL N = 0
SPHERE C = -.0882059039
T = .515714284
INDEX N = 1.4645 DEL N = 7.063E-03
SPHERE C = .0591170745
T = 2.61402069
INDEX N = 1.67003 DEL N = .014222
SPHERE C = -.164805104
T = .0608817272
AIR N = 1 DEL N = 0
SPHERE C = -.180423127
T = 1.88926659
INDEX N = 1.4645 DEL N = 7.063E-03
SPHERE C = -.162851163
T = .684318171
INDEX N = 1.9525 DEL N = .046774
SPHERE C = -.113479889
T = 32.4972984
AIR N = 1 DEL N = 0
IMAGE C = 0
F1 = -34.9088803 F2 = 31.6166678

TABLE 8.13

ZERO PETZVAL
FOCAL LENGTH: 30 CM F/8
SEMI FIELD ANGLE: 20 DEGREES

6 SURFACES

```
          MARGINAL        CHIEF
INITIAL RAYS
COORDS=    1.875000       0.000000
SLOPES=    0.000000                   3.639702E-1
    T =    .854001204     AIR
SURFACE #1 SPHERE    C = -.172295044       R = -5.80399747
COORDS=    1.875000      -3.108310E-1
INCDNT=   -3.230532E-1   -3.104156E-1
SLOPES=   -1.024638E-1    2.655148E-1
SPHAB  =-1.003075E-2  PETZ  = 5.464735E-2  COMA  =-9.638351E-3  CURVA =-2.790819E-2
ASTIG  =-9.261305E-3  DIST  =-2.681644E-2  LONCO = 2.921293E-3  LATCO = 2.807014E-3
    T =    .2640183          N = 1.4645 DEL N = 7.063E-03
SURFACE #2 SPHERE    C = .115475025       R = 8.65988122
COORDS=    1.902052      -3.809318E-1
INCDNT=    3.221033E-1   -3.095029E-1
SLOPES=   -6.282264E-2    2.274244E-1
SPHAB  = 1.003075E-2  PETZ  =-9.703974E-3  COMA  =-9.638351E-3  CURVA = 1.257252E-2
ASTIG  = 9.261305E-3  DIST  =-1.208069E-2  LONCO =-3.313682E-3  LATCO = 3.184053E-3
    T =    1.33823965        N = 1.67003 DEL N = .014222
SURFACE #3 SPHERE    C = -.321918391       R = -3.10637736
COORDS=    1.986124      -6.852801E-1
INCDNT=   -5.765472E-1   -6.820098E-3
SLOPES=    3.234813E-1    2.319940E-1
SPHAB  = 4.871400E-1  PETZ  =-1.291564E-1  COMA  = 5.762481E-3  CURVA = 4.413917E-2
ASTIG  = 6.816560010E-5      DIST  = 5.221315E-4  LONCO =-1.628553E-2  LATCO =-1.926450E-4
    T =    .0311682081    AIR
SURFACE #4 SPHERE    C = -.352425509       R = -2.83747905
COORDS=    1.976042      -6.925109E-1
INCDNT=   -1.019889       1.206448E-2
SLOPES=   -6.516530E-10   2.358206E-1
SPHAB  =-4.871400E-1  PETZ  = 1.117799E-1  COMA  = 5.762483E-3  CURVA =-3.820993E-2
ASTIG  =-6.816565E-5  DIST  = 4.519935E-4  LONCO = 9.719608E-3  LATCO =-1.149753E-4
    T =    4.68881004        N = 1.4645 DEL N = 7.063E-03
SURFACE #5 SPHERE    C = -.162851163       R = -6.14057635
COORDS=    1.976042      -1.798229
INCDNT=   -3.218007E-1    5.702308E-2
SLOPES=   -8.042957E-2    2.500727E-1
SPHAB  =-1.324575E-2  PETZ  = 2.779267E-2  COMA  = 2.347146E-3  CURVA =-9.899387E-3
ASTIG  =-4.159142E-4  DIST  = 1.754171E-3  LONCO = 1.781800E-2  LATCO =-3.157349E-3
    T =    .684318171        N = 1.9525 DEL N = .046774
SURFACE #6 SPHERE    C = -.113479889       R = -8.81213411
COORDS=    2.031081      -1.969358
INCDNT=   -1.500573E-1   -2.659015E-2
SLOPES=    6.250000E-2    2.753998E-1
SPHAB  = 1.324574E-2  PETZ  =-5.535959E-2  COMA  = 2.347146E-3  CURVA = 1.930583E-2
ASTIG  = 4.159142E-4  DIST  = 3.420993E-3  LONCO =-1.425571E-2  LATCO =-2.526112E-3
    T =    32.4972984    AIR
SURFACE #7 IMAGE
COORDS=    7.507697E-8   -1.091911E1
   SUMS OF ABERRATIONS
SPHAB  =-2.637535E-9  PETZ  = 6.402843E-10  COMA  =-3.057446E-3  CURVA =-2.692104E-10
ASTIG  =-5.127276E-11  DIST  =-3.274784E-2  LONCO =-3.396026E-3  LATCO =-1.355238E-8
```

TABLE 8.14

ZERO COMA
FOCAL LENGTH: 30 CM F/8
SEMI FIELD ANGLE: 20 DEGREES

6 SURFACES

```
            MARGINAL        CHIEF
INITIAL RAYS
COORDS=    1.875000       0.000000
SLOPES=    0.000000                  3.639702E-1
  T =     1.09742029      AIR
SURFACE #1 SPHERE   C = -.134078234      R = -7.45833213
COORDS=    1.875000      -3.994283E-1
INCDNT=   -2.513967E-1   -3.104156E-1
SLOPES=   -7.973627E-2    2.655148E-1
SPHAB   =-4.727045E-3  PETZ   = 4.252601E-2  COMA  =-5.836786E-3  CURVA  =-2.171787E-2
ASTIG   =-7.207053E-3  DIST   =-2.681644E-2  LONCO = 2.273320E-3  LATCO  = 2.807014E-3
  T =     .339272402        N = 1.4645  DEL N = 7.063E-03
SURFACE #2 SPHERE    C = .0898614788      R = 11.1282389
COORDS=    1.902052      -4.895102E-1
INCDNT=    2.506575E-1   -3.095029E-1
SLOPES=   -4.888794E-2    2.274244E-1
SPHAB   = 4.727045E-3  PETZ   =-7.551533E-3  COMA  =-5.836786E-3  CURVA  = 9.783803E-3
ASTIG   = 7.207053E-3  DIST   =-1.208069E-2  LONCO =-2.578673E-3  LATCO  = 3.184053E-3
  T =     1.71968298        N = 1.67003 DEL N = .014222
SURFACE #3 SPHERE    C = -.250513587      R = -3.99179946
COORDS=    1.986124      -8.806079E-1
INCDNT=   -4.486631E-1   -6.820099E-3
SLOPES=    2.517298E-1    2.319940E-1
SPHAB   = 2.295674E-1  PETZ   =-1.005081E-1  COMA  = 3.489640E-3  CURVA  = 3.434865E-2
ASTIG   = 5.304578E-5  DIST   = 5.221315E-4  LONCO =-1.267323E-2  LATCO  =-1.926450E-4
  T =     .0400521965      AIR
SURFACE #4 SPHERE    C = -.274253913      R = -3.64625609
COORDS=    1.976042      -8.898998E-1
INCDNT=   -7.936670010E-1 1.206448E-2
SLOPES=   -2.396519E-10   2.358206E-1
SPHAB   =-2.295674E-1  PETZ   = 8.698596E-2  COMA  = 3.489640E-3  CURVA  =-2.973458E-2
ASTIG   =-5.304580E-5  DIST   = 4.519934E-4  LONCO = 7.563699E-3  LATCO  =-1.149753E-4
  T =     3.85177988        N = 1.4645  DEL N = 7.063E-03
SURFACE #5 SPHERE    C = -.162851163      R = -6.14057635
COORDS=    1.976042      -1.798229
INCDNT=   -3.218007E-1    5.702308E-2
SLOPES=   -8.042957E-2    2.500727E-1
SPHAB   =-1.324575E-2  PETZ   = 2.779267E-2  COMA  = 2.347146E-3  CURVA  =-9.899387E-3
ASTIG   =-4.159142E-4  DIST   = 1.754171E-3  LONCO = 1.781800E-2  LATCO  =-3.157349E-3
  T =     .684318171        N = 1.9525  DEL N = .046774
SURFACE #6 SPHERE    C = -.113479889      R = -8.81213411
COORDS=    2.031081      -1.969358
INCDNT=   -1.500573E-1   -2.659015E-2
SLOPES=    6.250000E-2    2.753998E-1
SPHAB   = 1.324574E-2  PETZ   =-5.535959E-2  COMA  = 2.347146E-3  CURVA  = 1.930583E-2
ASTIG   = 4.159142E-4  DIST   = 3.420993E-3  LONCO =-1.425571E-2  LATCO  =-2.526112E-3
  T =     32.4972984       AIR
SURFACE #7 IMAGE
COORDS=    5.479887E-8   -1.091911E1
  SUMS OF ABERRATIONS
SPHAB   =-1.422450E-9  PETZ   =-6.114625E-3  COMA  = 2.382876E-10 CURVA  = 2.086445E-3
ASTIG   =-2.046363E-12 DIST   =-3.274784E-2  LONCO =-1.852601E-3  LATCO  =-1.354601E-8
```

TABLE 8.15

ZERO LONGITUDINAL COLOR
FOCAL LENGTH: 30 CM F/8
SEMI FIELD ANGLE: 20 DEGREES

6 SURFACES

```
              MARGINAL        CHIEF
INITIAL RAYS
COORDS=    1.875000        0.000000
SLOPES=    0.000000               3.639702E-1
   T =    1.66814429       AIR
SURFACE #1 SPHERE    C = -.0882059039      R = -11.3371096
COORDS=    1.875000       -6.071549E-1
INCDNT=   -1.653861E-1    -3.104156E-1
SLOPES=   -5.245601E-2     2.655148E-1
SPHAB     =-1.345885E-3   PETZ    = 2.797654E-2   COMA   =-2.526111E-3   CURVA  =-1.428751E-2
ASTIG     =-4.741296E-3   DIST    =-2.681644E-2   LONCO  = 1.495547E-3   LATCO  = 2.807014E-3
   T =     .515714284       N = 1.4645 DEL N = 7.063E-03
SURFACE #2 SPHERE    C = .0591170745      R = 16.9155867
COORDS=    1.902052       -7.440846E-1
INCDNT=    1.648998E-1    -3.095029E-1
SLOPES=   -3.216185E-2     2.274244E-1
SPHAB     = 1.345885E-3   PETZ    =-4.967919E-3   COMA   =-2.526111E-3   CURVA  = 6.436460E-3
ASTIG     = 4.741296E-3   DIST    =-1.208069E-2   LONCO  =-1.696429E-3   LATCO  = 3.184053E-3
   T =    2.61402069       N = 1.67003 DEL N = .014222
SURFACE #3 SPHERE    C = -.164805104      R = -6.06777324
COORDS=    1.986124       -1.338577
INCDNT=   -2.951615E-1    -6.820099E-3
SLOPES=    1.656052E-1     2.319940E-1
SPHAB     = 6.536246E-2   PETZ    =-6.612119E-2   COMA   = 1.510286E-3   CURVA  = 2.259691E-2
ASTIG     = 3.489717E-5   DIST    = 5.221315E-4   LONCO  =-8.337326E-3   LATCO  =-1.926450E-4
   T =     .0608817272     AIR
SURFACE #4 SPHERE    C = -.180423127      R = -5.54252671
COORDS=    1.976042       -1.352701
INCDNT=   -5.221289E-1     1.206448E-2
SLOPES=   -4.640697E-10    2.358206E-1
SPHAB     =-6.536246E-2   PETZ    = 5.722536E-2   COMA   = 1.510287E-3   CURVA  =-1.956146E-2
ASTIG     =-3.489718E-5   DIST    = 4.519934E-4   LONCO  = 4.975923E-3   LATCO  =-1.149753E-4
   T =    1.88926659       N = 1.4645 DEL N = 7.063E-03
SURFACE #5 SPHERE    C = -.162851163      R = -6.14057635
COORDS=    1.976042       -1.798229
INCDNT=   -3.218007E-1     5.702308E-2
SLOPES=   -8.042957E-2     2.500727E-1
SPHAB     =-1.324575E-2   PETZ    = 2.779267E-2   COMA   = 2.347146E-3   CURVA  =-9.899387E-3
ASTIG     =-4.159142E-4   DIST    = 1.754171E-3   LONCO  = 1.781800E-2   LATCO  =-3.157349E-3
   T =     .684318171       N = 1.9525 DEL N = .046774
SURFACE #6 SPHERE    C = -.113479889      R = -8.81213411
COORDS=    2.031081       -1.969358
INCDNT=   -1.500573E-1    -2.659015E-2
SLOPES=    6.250000E-2     2.753998E-1
SPHAB     = 1.324574E-2   PETZ    =-5.535959E-2   COMA   = 2.347146E-3   CURVA  = 1.930583E-2
ASTIG     = 4.159142E-4   DIST    = 3.420993E-3   LONCO  =-1.425571E-2   LATCO  =-2.526112E-3
   T =    32.4972984       AIR
SURFACE #7 IMAGE
COORDS=    6.614937E-8    -1.091911E1
  SUMS OF ABERRATIONS
SPHAB     =-9.022187010E-10      PETZ   =-1.345412E-2   COMA   = 2.662642E-3   CURVA  = 4.590843E-3
ASTIG     =-1.841727E-11  DIST    =-3.274784E-2   LONCO  = 1.818989E-10  LATCO  =-1.355602E-8
```

Table 8.16.

Focal length:15 cm   F/6.25
Format:10

SURFACE #1	T = 4.385626E1 PUPIL	AIR			
SURFACE #2	T =-8.222614 SPHERE	AIR C = 1.055134E-1 R = 9.477473			
SURFACE #3	T = 1.746941 SPHERE	INDEX C = 1.587261E-1 R = 6.300163	N = 1.952500	DEL N = 4.677400E-2
SURFACE #4	T = 6.188330E-1 SPHERE	INDEX C = 1.901655E-1 R = 5.258578	N = 1.464500	DEL N = 7.063000E-3
SURFACE #5	T = 2.579575 SPHERE	AIR C = 1.188865E-1 R = 8.411381			
SURFACE #6	T = 2.065839 SPHERE	INDEX C =-4.416416E-2 R =-2.264279E1	N = 1.952500	DEL N = 4.677400E-2
SURFACE #7	T = 1.671910 SPHERE	INDEX C = 9.870953E-2 R = 1.013073E1	N = 1.622300	DEL N = 1.170700E-2
SURFACE #8	T = 1.163620 STOP	AIR			
SURFACE #9	T = 7.410810E-1 SPHERE	AIR C =-1.976835E-1 R =-5.058592			
SURFACE #10	T = 2.423600E-1 SPHERE	INDEX C = 1.323368E-1 R ≐ 7.556477	N = 1.464500	DEL N = 7.063000E-3
SURFACE #11	T = 1.001328 SPHERE	INDEX C =-3.841937E-1 R =-2.602854	N = 1.670030	DEL N = 1.422200E-2
SURFACE #12	T = 1.203310E-1 SPHERE	AIR C =-4.332399E-1 R =-2.308190			
SURFACE #13	T = 2.123245 SPHERE	INDEX C =-2.719993E-1 R =-3.676480	N = 1.464500	DEL N = 7.063000E-3
SURFACE #14	T = 4.097180E-1 SPHERE	INDEX C =-1.895378E-1 R =-5.275994	N = 1.952500	DEL N = 4.677400E-2
SURFACE #15	T =-4.281387 PUPIL	AIR			
SURFACE #16	T = 2.373815E1 IMAGE	AIR			

Initial rays							Marginal	Chief
						COORDS	0.000000	1.000000E1
						SLOPES	-2.736212E-2	2.280176E-1
SURFACE #1 ′ PUPIL								
						COORDS	1.200000	0.000000
SURFACE #2 SPHERE								
SPHAB	1.386811E-3	PETZ	-5.147323E-2	LATCO	-3.042034E-3	COORDS	9.750119E-1	1.874901
COMA	-3.214750E-4	CURVA	7.116603E-3	LONGCO	7.051703E-4	INCIDS	1.302389E-1	-3.019053E-2
ASTIG	7.452073E-5	DIST	-1.649691E-3	LAGRA	-2.736212E-1	SLOPES	3.617312E-2	2.132896E-1
SURFACE #3 SPHERE								
SPHAB	-1.386811E-3	PETZ	2.708867E-2	LATCO	3.697787E-3	COORDS	9.118196E-1	1.502297
COMA	-3.214712E-4	CURVA	-3.780535E-3	LONGCO	8.571696E-4	INCIDS	1.085564E-1	2.516404E-2
ASTIG	-7.451898E-5	DIST	-8.763513E-4	LAGRA	-2.736212E-1	SLOPES	9.983978E-10	2.049045E-1

Table 8.16. (cont.)

SURFACE #4	SPHERE							
SPHAB	-8.653973E-3	PETZ	6.031538E-2	LATCO	1.116706E-3	COORDS	9.118196E-1	1.375495
COMA	-2.828180E-3	CURVA	-9.176052E-3	LONGCO	3.649474E-4	INCIDS	1.733966E-1	5.666725E-2
ASTIG	-9.242694E-4	DIST	-2.998799E-3	LAGRA	-2.736212E-1	SLOPES	-8.054273E-2	1.785825E-1
SURFACE #5	SPHERE							
SPHAB	8.653972E-3	PETZ	-5.799715E-2	LATCO	-5.730154E-3	COORDS	1.119586	9.148281E-1
COMA	-2.828205E-3	CURVA	8.858909E-3	LONGCO	1.872672E-3	INCIDS	2.136464E-1	-6.982178E-2
ASTIG	9.242857E-4	DIST	-2.895180E-3	LAGRA	-2.736212E-1	SLOPES	2.368169E-2	1.445209E-1
SURFACE #6	SPHERE							
SPHAB	4.271832E-4	PETZ	-4.603889E-3	LATCO	-2.483388E-3	COORDS	1.070663	6.162711E-1
COMA	1.033776E-3	CURVA	3.131582E-3	LONGCO	-6.009756E-3	INCIDS	-7.096663E-2	-1.717380E-1
ASTIG	2.501721E-3	DIST	7.578375E-3	LAGRA	-2.736212E-1	SLOPES	3.812611E-2	1.794762E-1
SURFACE #7	SPHERE							
SPHAB	-4.271832E-4	PETZ	3.786411E-2	LATCO	7.222094E-4	COORDS	1.006920	3.162031E-1
COMA	1.033777E-3	CURVA	-7.681937E-3	LONGCO	-1.747736E-3	INCIDS	6.126645E-2	-1.482639E-1
ASTIG	-2.501726E-3	DIST	1.859018E-2	LAGRA	-2.736212E-1	SLOPES	-8.081997E-10	2.717408E-1
SURFACE #8	STOP							
						COORDS	1.006920	-5.593392E-10
SURFACE #9	SPHERE							
SPHAB	-3.142805E-3	PETZ	6.269988E-2	LATCO	9.666288E-4	COORDS	1.006920	-2.013820E-1
COMA	-3.661938E-3	CURVA	-1.284483E-2	LONGCO	1.126298E-3	INCIDS	-1.990513E-1	-2.319310E-1
ASTIG	-4.266823E-3	DIST	-1.496656E-2	LAGRA	-2.736212E-1	SLOPES	-6.313373E-2	1.981786E-1
SURFACE #10	SPHERE							
SPHAB	3.142805E-3	PETZ	-1.112096E-2	LATCO	-1.096993E-3	COORDS	1.022221	-2.494125E-1
COMA	-3.661939E-3	CURVA	5.788289E-3	LONGCO	1.278196E-3	INCIDS	1.984111E-1	-2.311850E-1
ASTIG	4.266825E-3	DIST	-6.744409E-3	LAGRA	-2.736212E-1	SLOPES	-3.871535E-2	1.697267E-1
SURFACE #11	SPHERE							
SPHAB	1.704691E-1	PETZ	-1.541417E-1	LATCO	-5.566606E-3	COORDS	1.060987	-4.193647E-1
COMA	3.978350E-3	CURVA	2.118106E-2	LONGCO	-1.299116E-4	INCIDS	-3.689093E-1	-8.609480E-3
ASTIG	9.284539E-5	DIST	4.943165E-4	LAGRA	-2.736212E-1	SLOPES	2.084650E-1	1.754953E-1
SURFACE #12	SPHERE							
SPHAB	-1.704691E-1	PETZ	1.374120E-1	LATCO	3.283640E-3	COORDS	1.035903	-4.404822E-1
COMA	3.978409E-3	CURVA	-1.889227E-2	LONGCO	-7.663358E-5	INCIDS	-6.572593E-1	1.533912E-2
ASTIG	-9.284810E-5	DIST	4.409078E-4	LAGRA	-2.736212E-1	SLOPES	-1.113722E-8	1.803605E-1
SURFACE #13	SPHERE							
SPHAB	-1.162562E-2	PETZ	4.642022E-2	LATCO	8.178647E-3	COORDS	1.035903	-8.234317E-1
COMA	1.799446E-3	CURVA	-6.629301E-3	LONGCO	-1.265915E-3	INCIDS	-2.817648E-1	4.361237E-2
ASTIG	-2.785235E-4	DIST	1.026102E-3	LAGRA	-2.736212E-1	SLOPES	-7.042317E-2	1.912608E-1
SURFACE #14	SPHERE							
SPHAB	1.162561E-2	PETZ	-9.246336E-2	LATCO	-6.543522E-3	COORDS	1.064756	-9.017947E-1
COMA	1.799445E-3	CURVA	1.292849E-2	LONGCO	-1.012825E-3	INCIDS	-1.313883E-1	-2.033665E-2
ASTIG	2.785231E-4	DIST	2.001108E-3	LAGRA	-2.736212E-1	SLOPES	5.472422E-2	2.106314E-1
SURFACE #15	PUPIL							
						COORDS	1.299052	-2.202796E-9
SURFACE #16	IMAGE							
						COORDS	3.237528E-7	-5.000000

Sums of Aberrations								
SPHAB	-3.711466E-8	PETZ	7.566996E-10	LATCO	-6.497079E-3			
COMA	-4.849880E-9	CURVA	1.127046E-8	LONGCO	-4.038324E-3			
ASTIG	1.138312E-8	DIST	-8.512870E-10	LAGRA	-2.736212E-1			

Table 8.17.

Focal length:20 cm F/8				
Format:10				

SURFACE #1	T = 3.999999E1 PUPIL	AIR		
SURFACE #2	T =-8.185135 SPHERE	AIR C = 1.242499E-1 R = 8.048293		
SURFACE #3	T = 3.429107 SPHERE	INDEX C = 2.101924E-1 R = 4.757545	N = 1.952500	DEL N = 4.677400E-2
SURFACE #4	T = 8.045920E-1 SPHERE	INDEX C = 2.889175E-1 R = 3.461196	N = 1.464500	DEL N = 7.063000E-3
SURFACE #5	T = 5.618940E-1 SPHERE	AIR C = 2.373400E-1 R = 4.213364		
SURFACE #6	T = 1.147563 SPHERE	INDEX C =-7.296463E-2 R =-1.370527E1	N = 1.668820	DEL N = 1.165500E-2
SURFACE #7	T = 3.535680E-1 SPHERE	INDEX C = 1.082197E-1 R = 9.240464	N = 1.464500	DEL N = 7.063000E-3
SURFACE #8	T = 1.382670 STOP	AIR		
SURFACE #9	T = 1.382670 SPHERE	AIR C =-1.082197E-1 R =-9.240461		
SURFACE #10	T = 3.535680E-1 SPHERE	INDEX C = 7.296465E-2 R = 1.370527E1	N = 1.464500	DEL N = 7.063000E-3
SURFACE #11	T = 1.147563 SPHERE	INDEX C =-2.373401E-1 R =-4.213363	N = 1.668820	DEL N = 1.165500E-2
SURFACE #12	T = 5.618940E-1 SPHERE	AIR C =-2.889176E-1 R =-3.461195		
SURFACE #13	T = 8.045830E-1 SPHERE	INDEX C =-2.101924E-1 R =-4.757545	N = 1.464500	DEL N = 7.063000E-3
SURFACE #14	T = 3.429107 SPHERE	INDEX C =-1.242499E-1 R =-8.048293	N = 1.952500	DEL N = 4.677400E-2
SURFACE #15	T =-8.185136 PUPIL	AIR		
SURFACE #16	T = 3.999999E1 IMAGE	AIR		

		Marginal	Chief
Initial rays	COORDS SLOPES	0.000000 -3.125001E-2	1.000000E1 2.500000E-1
SURFACE #1 PUPIL	COORDS	1.250000	0.000000

Table 8.17. (cont.)

SURFACE #2	SPHERE							
SPHAB	2.054772E-3	PETZ	-6.061361E-2	LATCO	-3.686477E-3	COORDS	9.942145E-1	2.046284
COMA	5.642884E-5	CURVA	9.472428E-3	LONGCO	-1.012393E-4	INCIDS	1.547811E-1	4.250651E-3
ASTIG	1.549668E-6	DIST	2.601350E-4	LAGRA	-3.125001E-1	SLOPES	4.425780E-2	2.520737E-1
SURFACE #3	SPHERE							
SPHAB	-2.054772E-3	PETZ	3.587207E-2	LATCO	4.180044E-3	COORDS	8.424497E-1	1.181897
COMA	5.643561E-5	CURVA	-5.606563E-3	LONGCO	-1.148076E-4	INCIDS	1.328188E-1	-3.647951E-3
ASTIG	-1.550040E-6	DIST	1.639878E-4	LAGRA	-3.125001E-1	SLOPES	1.521016E-9	2.532892E-1
SURFACE #4	SPHERE							
SPHAB	-1.936341E-2	PETZ	9.163685E-2	LATCO	1.448275E-3	COORDS	8.424497E-1	9.781046E-1
COMA	-2.331124E-3	CURVA	-1.459890E-2	LONGCO	1.743550E-4	INCIDS	2.433984E-1	2.930227E-2
ASTIG	-2.806395E-4	DIST	-1.757534E-3	LAGRA	-3.125001E-1	SLOPES	-1.130586E-1	2.396783E-1
SURFACE #5	SPHERE							
SPHAB	1.936341E-2	PETZ	-9.511976E-2	LATCO	-2.075887E-3	COORDS	9.059767E-1	8.434308E-1
COMA	-2.331191E-3	CURVA	1.514312E-2	LONGCO	2.499193E-4	INCIDS	3.280831E-1	-3.949844E-2
ASTIG	2.806557E-4	DIST	-1.823104E-3	LAGRA	-3.125001E-1	SLOPES	1.842867E-2	2.238484E-1
SURFACE #6	SPHERE							
SPHAB	2.565232E-4	PETZ	-6.099924E-4	LATCO	-2.648395E-4	COORDS	8.848286E-2	5.865507E-1
COMA	8.242071E-4	CURVA	3.601285E-3	LONGCO	-8.509273E-4	INCIDS	-8.298986E-2	-2.666458E-1
ASTIG	2.648171E-3	DIST	1.157090E-2	LAGRA	-3.125001E-1	SLOPES	3.000702E-2	2.610495E-1
SURFACE #7	SPHERE							
SPHAB	-2.565232E-4	PETZ	3.432437E-2	LATCO	3.988840E-4	COORDS	8.742191E-1	4.942519E-1
COMA	8.242080E-4	CURVA	-8.011361E-3	LONGCO	-1.281613E-3	INCIDS	6.460068E-2	-2.075617E-1
ASTIG	-2.648177E-3	DIST	2.574047E-2	LAGRA	-3.125001E-1	SLOPES	-9.968630E-10	3.574620E-1
SURFACE #8	STOP							
						COORDS	8.742191E-1	-5.120455E-10
SURFACE #9	SPHERE							
SPHAB	-2.565234E-4	PETZ	3.432438E-2	LATCO	3.988841E-4	COORDS	8.742191E-1	-4.942519E-1
COMA	-8.242084E-4	CURVA	-8.011363E-3	LONGCO	1.281613E-3	INCIDS	-9.460773E-2	-3.039742E-1
ASTIG	-2.648178E-3	DIST	-2.574047E-2	LAGRA	-3.125001E-1	CLOPES	-3.000703E-2	2.610495E-1
SURFACE #10	SPHERE							
SPHAB	2.565234E-4	PETZ	-6.099926E-3	LATCO	-2.648395E-4	COORDS	8.848286E-2	-5.865507E-1
COMA	-8.242077E-4	CURVA	3.601286E-3	LONGCO	8.509273E-4	INCIDS	9.456824E-2	-3.038470E-1
ASTIG	2.648173E-3	DIST	-1.157090E-2	LAGRA	-3.125001E-1	SLOPES	-1.842868E-2	2.238484E-1
SURFACE #11	SPHERE							
SPHAB	1.936343E-2	PETZ	-9.511979E-2	LATCO	-2.075888E-3	COORDS	9.059767E-1	-8.434308E-1
COMA	2.331186E-3	CURVA	1.514313E-2	LONGCO	-2.499186E-4	INCIDS	-1.965959E-1	-2.366842E-2
ASTIG	2.806542E-4	DIST	1.823099E-3	LAGRA	-3.125001E-1	SLOPES	1.130586E-1	2.396783E-1
SURFACE #12	SPHERE							
SPHAB	-1.936343E-2	PETZ	9.163688E-2	LATCO	1.448275E-3	COORDS	8.424497E-1	-9.781046E-1
COMA	2.331132E-3	CURVA	-1.459891E-2	LONGCO	-1.743556E-4	INCIDS	-3.564571E-1	4.291331E-2
ASTIG	-2.806413E-4	DIST	1.757539E-3	LAGRA	-3.125001E-1	SLOPES	-8.160441E-10	2.532892E-1
SURFACE #13	SPHERE							
SPHAB	-2.054772E-3	PETZ	3.587207E-2	LATCO	4.180044E-3	COORDS	8.424497E-1	-1.181897
COMA	-5.643503E-5	CURVA	-5.606563E-3	LONGCO	1.148064E-4	INCIDS	-1.770766E-1	-4.863470E-3
ASTIG	-1.550008E-6	DIST	-1.539862E-4	LAGRA	-3.125001E-1	SLOPES	-4.425780E-2	2.520737E-1
SURFACE #14	SPHERE							
SPHAB	2.054772E-3	PETZ	-6.061361E-2	LATCO	-3.686477E-3	COORDS	9.942145E-1	-2.046284
COMA	-5.642942E-5	CURVA	9.472428E-3	LONGCO	1.012403E-4	INCIDS	-7.927329E-2	2.177052E-3
ASTIG	1.549699E-6	DIST	-2.601376E-4	LAGRA	-3.125001E-1	SLOPES	3.125001E-2	2.500000E-1
SURFACE #15	PUPIL							
						COORDS	1.250000	0.000000
SURFACE #16	IMAGE							
						COORDS	-2.572415E-8	-1.000000E1

Sums of Aberrations						
SPHAB	-1.424269E-9	PETZ	-4.816684E-9	LATCO	-1.109584E-10
COMA	2.569706E-9	CURVA	1.882290E-8	LONGCO	1.037392E-11
ASTIG	1.808372E-8	DIST	-2.548859E-10	LAGRA	-3.125001E-1

Table 8.18.

| Focal length:20 cm   F/8 |
| Format: 10 |

| | T = 3.333332E1 | AIR | | |
| SURFACE #1 | PUPIL | | | |

	T =-6.820943	AIR		
SURFACE #2	SPHERE	C = 1.490999E-1		
		R = 6.706911		

	T = 2.857589	INDEX	N = 1.952500	DEL N = 4.677400E-2
SURFACE #3	SPHERE	C = 2.522309E-1		
		R = 3.964621		

	T = 5.559000E-2	INDEX	N = 1.464500	DEL N = 7.006300E-3
SURFACE #4	SPHERE	C = 2.910814E-1		
		R = 3.435465		

	T = 5.577170E-1	AIR		
SURFACE #5	SPHERE	C = 2.391177E-1		
		R = 4.182041		

	T = 1.139032	INDEX	N = 1.668820	DEL N = 1.165500E-2
SURFACE #6	SPHERE	C =-7.351112E-2		
		R =-1.360338E1		

	T = 3.509400E-1	INDEX	N = 1.464500	DEL N = 7.063000E-3
SURFACE #7	SPHERE	C = 1.090302E-1		
		R = 9.171769		

| | T = 1.372390 | AIR | | |
| SURFACE #8 | STOP | | | |

	T = 1.393110	AIR		
SURFACE #9	SPHERE	C =-1.074092E-1		
		R =-9.310193		

	T = 3.562370E-1	INDEX	N = 1.464500	DEL N = 7.063000E-3
SURFACE #10	SPHERE	C = 7.241816E-2		
		R = 1.380869E1		

	T = 1.156222	INDEX	N = 1.668820	DEL N = 1.165500E-2
SURFACE #11	SPHERE	C =-2.355625E-1		
		R =-4.245158		

	T = 5.661340E-1	AIR		
SURFACE #12	SPHERE	C =-2.867536E-1		
		R =-3.487314		

	T = 1.941992	INDEX	N = 1.464500	DEL N = 7.063000E-3
SURFACE #13	SPHERE	C =-1.681539E-1		
		R =-5.946931		

	T = 4.286383	INDEX	N = 1.952500	DEL N = 4.677400E-2
SURFACE #14	SPHERE	C =-9.939996E-2		
		R =-1.006037E1		

| | T =-1.023143E1 | AIR | | |
| SURFACE #15 | PUPIL | | | |

| | T = 5.000000E1 | AIR | | |
| SURFACE #16 | IMAGE | | | |

							Marginal	Chief
Initial rays								
						COORDS	0.000000	1.000000E1
						SLOPES	-3.750001E-2	3.000001E-1
SURFACE #1	PUPIL							
						COORDS	1.250000	0.000000
SURFACE #2	SPHERE							
SPHAB	2.958872E-3	PETZ	-7.273633E-2	LATCO	-4.423773E-3	COORDS	9.942146E-1	2.046284
COMA	8.125577E-5	CURVA	1.364030E-2	LONGCO	-1.214845E-4	INCIDS	1.857373E-1	5.100669E-3
ASTIG	2.231424E-6	DIST	3.745863E-4	LAGRA	-3.750001E-1	SLOPES	5.310937E-2	3.024884E-1
SURFACE #3	SPHERE							
SPHAB	-2.958872E-3	PETZ	4.304649E-2	LATCO	5.026204E-3	COORDS	8.424498E-1	1.181896
COMA	8.126890E-5	CURVA	-8.073451E-3	LONGCO	-1.380506E-4	INCIDS	1.593825E-1	-4.377628E-3
ASTIG	-2.232146E-6	DIST	2.217468E-4	LAGRA	-3.750001E-1	SLOPES	8.876668E-10	3.039471E-1
SURFACE #4	SPHERE							
SPHAB	-1.650147E-2	PETZ	9.232319E-2	LATCO	1.447409E-3	COORDS	8.424498E-1	1.165000
COMA	-2.366170E-3	CURVA	-1.764989E-2	LONGCO	2.075461E-4	INCIDS	2.452215E-1	3.516267E-2
ASTIG	-3.392887E-4	DIST	-2.530844E-3	LAGRA	-3.750001E-1	SLOPES	-1.139054E-1	2.876140E-1

Table 8.18. (cont.)

SURFACE #5	SPHERE							
SPHAB	1.650147E-2	PETZ	-9.583219E-2	LATCO	-2.091435E-3	COORDS	9.059768E-1	1.004592
COMA	-2.366247E-3	CURVA	1.830785E-2	LONGCO	2.999037E-4	INCIDS	3.305404E-1	-4.739822E-2
ASTIG	3.393106E-4	DIST	-2.625275E-3	LAGRA	-3.750001E-1	SLOPES	1.856670E-2	2.686180E-1

SURFACE #6	SPHERE							
SPHAB	2.186087E-4	PETZ	-6.145611E-3	LATCO	-2.668231E-4	COORDS	8.848287E-1	6.986279E-1
COMA	8.365998E-4	CURVA	4.353909E-3	LONGCO	-1.021113E-3	INCIDS	-8.361145E-2	-3.199750E-1
ASTIG	3.201607E-3	DIST	1.666210E-2	LAGRA	-3.750001E-1	SLOPES	3.023177E-2	3.132594E-1

SURFACE #7	SPHERE							
SPHAB	-2.186087E-4	PETZ	3.458145E-2	LATCO	4.018716E-4	COORDS	8.742192E-1	5.886927E-1
COMA	8.366007E-4	CURVA	-9.685639E-3	LONGCO	-1.537936E-3	INCIDS	6.508454E-2	-2.490741E-1
ASTIG	-3.201615E-3	DIST	3.706629E-2	LAGRA	-3.750001E-1	SLOPES	1.316323E-9	4.289544E-1

SURFACE #8	STOP							
						COORDS	8.742192E-1	0.000000

SURFACE #9	SPHERE							
SPHAB	-2.090022E-4	PETZ	3.406730E-2	LATCO	3.958966E-4	COORDS	8.742192E-1	-5.975806E-1
COMA	-8.119079E-4	CURVA	-9.541628E-3	LONGCO	1.537934E-3	INCIDS	-9.389915E-2	-3.647687E-1
ASTIG	-3.154008E-3	DIST	-3.706623E-2	LAGRA	-3.750001E-1	SLOPES	-2.978228E-2	3.132595E-1

SURFACE #10	SPHERE							
SPHAB	2.090022E-4	PETZ	-6.054238E-3	LATCO	-2.628560E-4	COORDS	8.848287E-1	-7.091752E-1
COMA	-8.119084E-4	CURVA	4.289182E-3	LONGCO	1.021114E-3	INCIDS	9.385995E-2	-3.646167E-1
ASTIG	3.154012E-3	DIST	-1.666214E-2	LAGRA	-3.750001E-1	SLOPES	-1.829065E-2	2.686181E-1

SURFACE #11	SPHERE							
SPHAB	1.577633E-2	PETZ	-9.440737E-2	LATCO	-2.060340E-3	COORDS	9.059768E-1	-1.019757
COMA	2.296350E-3	CURVA	1.803564E-2	LONGCO	-2.998963E-4	INCIDS	-1.951235E-1	-2.840153E-2
ASTIG	3.342491E-4	DIST	2.625207E-3	LAGRA	-3.750001E-1	SLOPES	1.122118E-1	2.876136E-1

SURFACE #12	SPHERE							
SPHAB	-1.577633E-2	PETZ	9.095054E-2	LATCO	1.437428E-3	COORDS	8.424498E-1	-1.182585
COMA	2.296390E-3	CURVA	-1.738749E-2	LONGCO	-2.092309E-4	INCIDS	-3.537874E-1	5.149699E-2
ASTIG	-3.342605E-4	DIST	2.530909E-3	LAGRA	-3.750001E-1	SLOPES	-1.457124E-9	3.039471E-1

SURFACE #13	SPHERE							
SPHAB	-8.767030E-4	PETZ	2.869766E-2	LATCO	3.344036E-3	COORDS	8.424498E-1	-1.772848
COMA	-3.611545E-5	CURVA	-5.382300E-3	LONGCO	1.377563E-4	INCIDS	-1.416613E-1	-5.835682E-3
ASTIG	-1.487762E-6	DIST	-2.217218E-4	LAGRA	-3.750001E-1	SLOPES	-3.540625E-2	3.024885E-1

SURFACE #14	SPHERE							
SPHAB	8.767029E-4	PETZ	-4.849089E-2	LATCO	-2.949182E-3	COORDS	9.942146E-1	-3.069430
COMA	-3.611525E-5	CURVA	9.093532E-3	LONGCO	1.214974E-4	INCIDS	-6.341864E-2	2.612657E-3
ASTIG	1.487934E-6	DIST	-3.746263E-4	LAGRA	-3.750001E-1	SLOPES	2.500001E-2	3.000000E-1

SURFACE #15	PUPIL							
						COORDS	1.250000	1.274020E-8

SURFACE #16	İMAGE							
						COORDS	9.850010E-8	-1.500000E1

Sums of Aberrations

SPHAB	-1.609578E-9	PETZ	-4.336471E-9	LATCO	-1.563297E-6
COMA	-1.517321E-9	CURVA	6.792106E-9	LONGCO	-1.959066E-6
ASTIG	5.973784E-9	DIST	-6.116352E-10	LAGRA	-3.750001E-1

Table 8.19.

	T = INFINITY			
	T = -1.211728E1	AIR		
SURFACE #1	SPHERE	C = 6.276069E-2		
		R = 1.593354E1		
	T = 2.136093	INDEX	N = 1.808020	DEL N = 1.984100E-2
SURFACE #2	SPHERE	C = 1.186793E-1		
		R = 8.426068		
	T = 4.435290	INDEX	N = 1.670030	DEL N = 1.422200E-2
SURFACE #3	SPHERE	C = -7.890326E-2		
		R = -1.267375E1		
	T = 3.345730	INDEX	N = 1.603110	DEL N = 9.952000E-3
SURFACE #4	SPhERE	C = 8.642746E-2		
		R = 1.157040E1		
	T = 2.500000	AIR		
SURFACE #5	STOP			
	T = 2.060040	AIR		
SURFACE #6	SPHERE	C = -8.854076E-2		
		R = -1.129423E1		

Table 8.19. (cont.)

	T = 2.346833	INDEX	N = 1.578450	DEL N = 1.390800E-2
SURFACE #7	SPHERE	C = 6.428826E-2		
		R = 1.555494E1		

	T = 2.810943	INDEX	N = 1.701810	DEL N = 1.711200E-2
SURFACE #8	SPHERE	C = 1.740546E-1		
		R = 5.745324		

	T = 2.899046	INDEX	N = 1.696800	DEL N = 1.257500E-2
SURFACE #9	SPHERE	C =-5.133070E-2		
		R =-1.948152E1		

	T =-8.949513	AIR		
SURFACE #10	IMAGE			

Initial rays						Marginal	Chief
					COORDS	2.000000	0.000000
					SLOPES	0.000000	1.316525E-1

SURFACE #1	SPHERE							
SPHAB	1.856563E-3	PETZ	-2.804830E-2	LATCO	-2.754914E-3	COORDS	2.000000	1.595270
COMA	-4.663873E-4	CURVA	3.809790E-3	LONGCO	6.920622E-4	INCIDS	1.255214E-1	-3.153224E-2
ASTIG	1.171612E-4	DIST	-9.570578E-4	LAGRA	-2.633050E-1	SLOPES	5.609660E-2	1.175605E-1

SURFACE #2	SPHERE							
SPHAB	-1.856563E-3	PETZ	5.423694E-3	LATCO	1.395669E-3	COORDS	1.880172	1.344150
COMA	-4.663872E-4	CURVA	-8.312040E-4	LONGCO	3.506062E-4	INCIDS	1.670410E-1	4.196237E-2
ASTIG	-1.171611E-4	DIST	-2.088068E-4	LAGRA	-2.633050E-1	SLOPES	4.229446E-2	1.140932E-1

SURFACE #3	SPHERE							
SPHAB	1.562231E-3	PETZ	-1.972256E-3	LATCO	-1.147243E-3	COORDS	1.692584	8.381137E-1
COMA	1.601128E-3	CURVA	1.900646E-3	LONGCO	-1.175807E-3	INCIDS	-1.758449E-1	-1.802231E-1
ASTIG	1.640993E-3	DIST	1.947969E-3	LAGRA	-2.633050E-1	SLOPES	4.963491E-2	1.216164E-1

SURFACE #4	SPHERE							
SPHAB	-1.562231E-3	PETZ	3.251509E-2	LATCO	1.250269E-3	COORDS	1.526519	4.312179E-1
COMA	1.601128E-3	CURVA	-5.921686E-3	LONGCO	-1.281398E-3	INCIDS	8.229827E-2	-8.434737E-2
ASTIG	-1.640993E-3	DIST	6.069127E-3	LAGRA	-2.633050E-1	SLOPES	-1.833428E-9	1.724872E-1

SURFACE #5	STOP							
						COORDS	1.526519	-4.688627E-8

SURFACE #6	SPHERE							
SPHAB	-1.661705E-3	PETZ	3.244727E-2	LATCO	1.817949E-3	COORDS	1.526519	-3.553305E-1
COMA	-1.733833E-3	CURVA	-6.080858E-3	LONGCO	1.896859E-3	INCIDS	-1.351592E-1	-1.410259E-1
ASTIG	-1.809093E-3	DIST	-6.344806E-3	LAGRA	-2.633050E-1	SLOPES	-4.953139E-2	1.208058E-1

SURFACE #7	SPHERE							
SPHAB	1.661705E-3	PETZ	-2.952325E-3	LATCO	-5.004427E-4	COORDS	1.642761	-6.388416E-1
COMA	-1.733834E-3	CURVA	2.197774E-3	LONGCO	5.221653E-4	INCIDS	1.551417E-1	-1.618758E-1
ASTIG	1.809093E-3	DIST	-2.293172E-3	LAGRA	-2.633050E-1	SLOPES	-3.828555E-2	1.090718E-1

SURFACE #8	SPHERE							
SPHAB	-7.508583E-4	PETZ	3.019823E-4	LATCO	2.701217E-3	COORDS	1.750380	-9.454363E-1
COMA	5.990919E-4	CURVA	-5.177578E-4	LONGCO	-2.155237E-3	INCIDS	3.429472E-1	-2.736294E-1
ASTIG	-4.780011E-4	DIST	4.131066E-4	LAGRA	-2.633050E-1	SLOPES	-3.929815E-2	1.098798E-1

SURFACE #9	SPHERE							
SPHAB	7.508597E-4	PETZ	-2.107923E-2	LATCO	-1.322176E-3	COORDS	1.864307	-1.263983
COMA	5.990930E-4	CURVA	3.253135E-3	LONGCO	-1.054933E-3	INCIDS	-5.639803E-2	-4.499863E-2
ASTIG	4.780019E-4	DIST	2.595598E-3	LAGRA	-2.633050E-1	SLOPES	-7.865992E-10	1.412348E-1

SURFACE #10	IMAGE							
						COORDS	1.864307	-6.951086E-8

Sums of Aberrations

SPHAB	1.065018E-9	PETZ	1.663593E-2	LATCO	1.440329E-3
COMA	-9.879386E-10	CURVA	-2.190161E-3	LONGCO	-2.205682E-3
ASTIG	1.096396E-9	DIST	1.221957E-3	LAGRA	-2.633050E-1

Table 8.20.

	T = INFINITY			
	T =-7.270369	AIR		
SURFACE #1	SPHERE	C = 1.046011E-1		
		R = 9.560125		

	T = 1.281656	INDEX	N = 1.808020	DEL N = 1.984100E-2
SURFACE #2	SPHERE	C = 1.977989E-1		
		R = 5.055641		

182

Table 8.20. (cont.)

SURFACE #3	T = 2.661174 SPHERE	INDEX C =-1.315054E-1 R =-7.604248	N = 1.670030	DEL N = 1.422200E-2
SURFACE #4	T = 2.007438 SPHERE	INDEX C = 1.440458E-1 R = 6.942238	N = 1.603110	DEL N = 9.952000E-3
SURFACE #5	T = 1.500000 STOP	AIR		
SURFACE #6	T = 2.653200 SPHERE	AIR C =-6.874628E-2 R =-1.454624E1		
SURFACE #7	T = 3.022569 SPHERE	INDEX C = 4.991576E-2 R = 2.003375E1	N = 1.578450	DEL N = 1.390800E-2
SURFACE #8	T = 3.620313 SPHERE	INDEX C = 1.351424E-1 R = 7.399605	N = 1.701810	DEL N = 1.711200E-2
SURFACE #9	T = 3.733783 SPHERE	INDEX C =-3.985503E-2 R =-2.509093E1	N = 1.696800	DEL N = 1.257500E-2
SURFACE #10	T =-1.152639E1 IMAGE	AIR		

						Marginal	Chief
Initial rays							
					COORDS SLOPES	2.000000 0.000000	0.000000 1.316525E-1
SURFACE #1 SPHERE							
SPHAB	8.595198E-3	PETZ	-4.674717E-2	LATCO	-4.591523E-3	COORDS	2.000000	9.571622E-1
COMA	-1.295520E-3	CURVA	6.349650E-3	LONGCO	6.920620E-4	INCIDS	2.092023E-1	-3.153223E-2
ASTIG	1.952686E-4	DIST	-9.570575E-4	LAGRA	-2.633050E-1	SLOPES	9.349434E-2	1.175605E-1
SURFACE #2 SPHERE							
SPHAB	-8.595198E-3	PETZ	9.039490E-3	LATCO	2.326116E-3	COORDS	1.880172	8.064902E-1
COMA	-1.295520E-3	CURVA	-1.385340E-3	LONGCO	3.506062E-4	INCIDS	2.784016E-1	4.196237E-2
ASTIG	-1.952686E-4	DIST	-2.088068E-4	LAGRA	-2.633050E-1	SLOPES	7.049078E-2	1.140932E-1
SURFACE #3 SPHERE							
SPHAB	7.232550E-3	PETZ	-3.287094E-3	LATCO	-1.912072E-3	COORDS	1.692584	6.028682E-1
COMA	4.447577E-3	CURVA	3.167743E-3	LONGCO	-1.175807E-3	INCIDS	-2.930748E-1	-1.802231E-1
ASTIG	2.734989E-3	DIST	1.947969E-3	LAGRA	-2.633050E-1	SLOPES	8.272485E-2	1.216164E-1
SURFACE #4 SPHERE							
SPHAB	-7.232549E-3	PETZ	5.419182E-2	LATCO	2.083781E-3	COORDS	1.526519	2.587308E-1
COMA	4.447577E-3	CURVA	-9.869477E-3	LONGCO	-1.281398E-3	INCIDS	1.371638E-1	-8.434737E-2
ASTIG	-2.734989E-3	DIST	6.069127E-3	LAGRA	-2.633050E-1	SLOPES	3.175501E-9	1.724872E-1
SURFACE #5 STOP							
						COORDS	1.526519	-2.543675E-8
SURFACE #6 SPHERE							
SPHAB	-7.778069E-4	PETZ	2.519325E-2	LATCO	1.411522E-3	COORDS	1.526519	-4.576430E-1
COMA	-1.045248E-3	CURVA	-4.721400E-3	LONGCO	1.896859E-3	INCIDS	-1.049425E-1	-1.410259E-1
ASTIG	-1.404646E-3	DIST	-6.344805E-3	LAGRA	-2.633050E-1	SLOPES	-3.845798E-2	1.208058E-1
SURFACE #7 SPHERE							
SPHAB	7.778066E-4	PETZ	-2.292293E-3	LATCO	-3.885620E-4	COORDS	1.642761	-8.227869E-1
COMA	-1.045248E-3	CURVA	1.706433E-3	LONGCO	5.221653E-4	INCIDS	1.204576E-1	-1.618759E-1
ASTIG	1.404647E-3	DIST	-2.293173E-3	LAGRA	-2.633050E-1	SLOPES	-2.972630E-2	1.090718E-1
SURFACE #8 SPHERE							
SPHAB	-3.514599E-4	PETZ	2.344701E-4	LATCO	2.097324E-3	COORDS	1.750380	-1.217661
COMA	3.611648E-4	CURVA	-4.020063E-4	LONGCO	-2.155237E-3	INCIDS	2.662767E-1	-2.736294E-1
ASTIG	-3.711377E-4	DIST	4.131069E-4	LAGRA	-2.633050E-1	SLOPES	-3.051251E-2	1.098798E-1
SURFACE #9 SPHERE							
SPHAB	3.514607E-4	PETZ	-1.636668E-2	LATCO	-1.026586E-3	COORDS	1.864307	-1.627928
COMA	3.611654E-4	CURVA	2.525853E-3	LONGCO	-1.054932E-3	INCIDS	-4.378949E-2	-4.499862E-2
ASTIG	3.711380E-4	DIST	2.595598E-3	LAGRA	-2.633050E-1	SLOPES	3.248147E-9	1.412348E-1
SURFACE #10 IMAGE							
						COORDS	1.864307	-4.771537E-7

Sums of Aberrations					
SPHAB	1.679837E-9	PETZ	1.996579E-2	LATCO	-7.730705E-11
COMA	4.935948E-3	CURVA	-2.628545E-3	LONGCO	-2.205683E-3
ASTIG	1.214403E-9	DIST	1.221958E-3	LAGRA	-2.633050E-1

Table 8.21.

SURFACE #1	T = INFINITY T =-1.136922E1 SPHERE	AIR C = 7.234332E-2 R = 1.382298E1					
SURFACE #2	T = 3.890423 SPHERE	INDEX C = 1.491121E-1 R = 6.706364	N = 1.808020	DEL N = 1.984100E-2			
SURFACE #3	T = 2.023957 SPHERE	INDEX C =-9.203122E-2 R =-1.086588E1	N = 1.668820	DEL N = 1.165500E-2			
SURFACE #4	T = 3.360656 SPHERE	INDEX C = 1.028402E-1 R = 9.723820	N = 1.603110	DEL N = 9.952000E-3			
SURFACE #5	T = 2.000000 IMAGE	AIR					

Initial rays						Marginal	Chief
					COORDS SLOPES	2.000000 0.000000	0.000000 1.316525E-1
SURFACE #1 SPHERE SPHAB 2.843425E-3 COMA -4.592755E-4 ASTIG 7.418307E-5	PETZ -3.233086E-2 CURVA 4.330622E-3 DIST -6.994904E-4		LATCO -3.175548E-3 LONGCO 5.129207E-4 LAGRA -2.633050E-1		COORDS INCIDS SLOPES	2.000000 1.446866E-1 6.466173E-2	1.496786 -2.337006E-2 1.212082E-1
SURFACE #2 SPHERE SPHAB -2.843425E-3 COMA -4.592754E-4 ASTIG -7.418304E-5	PETZ 6.879223E-3 CURVA -9.798499E-4 DIST -1.582672E-4		LATCO 2.472787E-3 LONGCO 3.994092E-4 LAGRA -2.633050E-1		COORDS INCIDS SLOPES	1.748439 1.960516E-1 4.830863E-2	1.025234 3.166662E-2 1.185668E-1
SURFACE #3 SPHERE SPHAB 2.206762E-3 COMA 2.103314E-3 ASTIG 2.004715E-3	PETZ -2.260444E-3 CURVA 2.302308E-3 DIST 2.194380E-3		LATCO -4.280202E-4 LONGCO -4.079555E-4 LAGRA -2.633050E-1		COORDS INCIDS SLOPES	1.650664 -2.002212E-1 5.651551E-2	7.852602E-1 -1.908353E-1 1.263890E-1
SURFACE #4 SPHERE SPHAB -2.206762E-3 COMA 2.103313E-3 ASTIG -2.004715E-3	PETZ 3.868978E-2 CURVA -7.098321E-3 DIST 6.765567E-3		LATCO 1.362238E-3 LONGCO -1.298379E-3 LAGRA -2.633050E-1		COORDS INCIDS SLOPES	1.460735 9.370680E-2 2.000093E-9	3.605103E-1 -8.931403E-2 1.802552E-1
SURFACE #5 IMAGE					COORDS	1.460735	-7.846393E-8

Sums of Aberrations

SPHAB	3.656169E-10	PETZ	1.097770E-2	LATCO	2.314557E-4		
COMA	3.288076E-3	CURVA	-1.445241E-3	LONGCO	-7.940044E-4		
ASTIG	2.819434E-11	DIST	8.102190E-3	LAGRA	-2.633050E-1		

9. Conclusion

Thus ends this dissertation. It is fitting at this point to attempt a critique of this work, to indicate where we may have failed and where we may have exceeded the mark.

What we have here is, for the first time ever, a *constructive* procedure for optical design. Beginning with prescribed values for the first-order parameters of the lens to be designed, and with a list of the third- and fifth-order aberrations which are to be made equal to zero, by following a sequence of well-defined operations we can arrive at an appropriate lens design. The path may be tortuous and convoled. Indeed, in practice, regressions may occur; one occasionally must backtrack to a previous step to alter the value of a previously assigned parameter. However, there is no massive feedback loop; no evaluations of tentative designs, followed by adjustments of parameters, followed by an iteration of the entire process. It is more like picking one's way through an elaborate, multidimensional maze in which regressions are usually the result of encountering a dead end.

One facet of the modular approach to optical design, one that never occurred to us until quite late in the game, was that it may lead to distinctly different configurations; configurations that would not necessarily be reached by a more conventional approach. Time and experience alone will tell whether the module will bear this kind of fruit.

I would extend the theory to include conic surfaces or other classes of aspheric surfaces and, perhaps, to consider finite-conjugate modules. One large area, demanding a thorough-going study, is the matter of selecting glasses. There are two approaches to be considered here. One deals with the relationship between the critical values of the module. The location of a module in a design determines in which category it should lie. Presumably, the first medium in that module is determined by the last glass in the next-preceding module. In principle, the other glasses could be chosen to enlarge the parameter domain of the appropriate category and, perhaps, control *a priori*, certain quantities that govern sufficient conditions for coupling.

A rather exciting area has to do with HERZBERGER's [9.1] idea of the super-achromat. He discovered an empirical relationship between the refractive index and the dispersion of optical glasses and showed that, in principle, it is possible to select glasses for use in a triplet design to correct the chromatic aberrations not only at isolated values of wavelength but over an entire wavelength range. STEVENS [9.2,3] conducted some experiments with prisms that indicated that the idea was practicable. Rumor has it that a triplet was indeed designed and a prototype manufactured and, also, that it was totally unacceptable. It is not said whether the superachromat principle was demonstrated.

It appears that the module is an ideal place to test and apply HERZBERGER's principle. The triplet design perhaps possesses too-few degrees of freedom to permit the kind of experimentation that is required. A design based on modules would be less apt to suffer in this way; indeed, with all of the degrees of freedom intrinsic in a modular design, a useful application of HERZBERGER's principle would appear to be inevitable.

Perhaps the most important step that should be made is to relax the requirement that the various aberrations should be set equal to zero. There is a technique in operations research known as the simplex method, by means of which a system of equations, which may be overdetermined or underdetermined, can be brought to a reasonable compromise solution. The application of this method to the lens-design equations would result in the control of more aberrations without the unnecessary and rather absurd idea that they all be made exactly equal to zero.

The design of catadioptric systems also appears to be a natural direction for research in modules. One possibility is shown in Fig.9.1. Note that it is a lens with two surfaces but acts like a system with four. What we have is a pair of modules, the first of which is in a forward orientation, the second in a backward orientation, coupled in the easy-way mode, to form a finite-

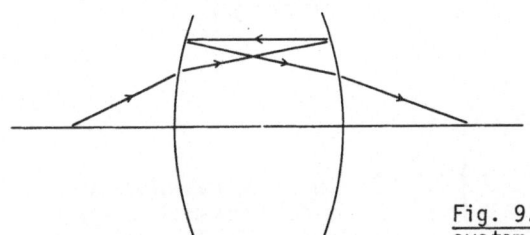

Fig. 9.1. Catadioptric finite-conjugate system

conjugate system. The first surface of the system is the first surface of the first module and the second surface of the second module. The system's second surface is the second surface of the first module and the first surface of the second module. The equations describing the system are

$$N_0^a = n_0 \quad , \quad N_1^a = n_1 \quad , \quad N_2^a = -n_1$$

$$N_0^b = n_2 \quad , \quad N_1^b = n_1 \quad , \quad N_2^b = -n_1 \tag{9.1}$$

$$c_1 = C_1^a/f_a = -C_2^b/f_b \qquad t_1 = f_a T_1^a = f_b T_1^b = f_a A^a + f_b A^b$$

$$c_2 = C_2^a/f_a = -C_1^b/f_b \tag{9.2}$$

The equations in (9.2) are four in number and contain four unknowns, two power parameters and two shape parameters. In principle, these are solvable in terms of the refractive index.

Something similar could be done with a pair of hard-way coupled modules which would form an afocal system as shown in Fig.9.2. However the hard-way coupling would add an additional equation to the system and would then require an additional independent variable. The only remaining uncommitted variable is the index of refraction. It may be that the value of N so obtained lies in the narrow region into which glasses fall. In that case such a device is possible; otherwise, not.

MERCADO [9.4] has studied catadioptric systems in which all-refracting and all-reflecting modules are mixed. An example of such a hybrid system is shown

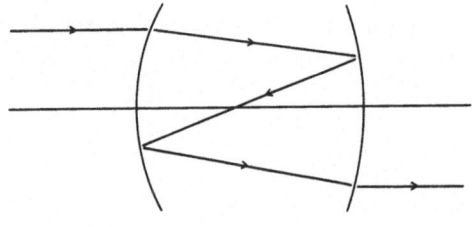

Fig. 9.2. Catadioptric afocal system

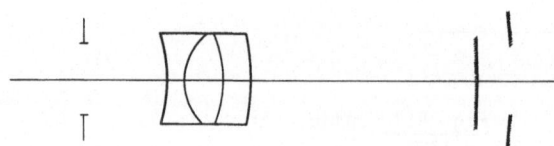

Fig. 9.3. Maksutov-Cassegrain three-module system. After MERCADO [9.4]

Table 9.1. Maksutov-Cassegrain 3-module system. Focal length = 100 mm, F/10; semi-field angle = 2 degrees;F1 = -90.9934; F2 = 103.785; F3 = 114.057; PSI-1 = -0.0334872; PSI-2 = 0.0939; PHI-3 = 126 degrees

---

Entrance pupil

T = 16.84	N = 1	DEL N = 0	Air	
Surface No. 1		C = -0.0222254		R = -44.9935
T = 3.82933	N = 1.51112	DEL N = 0.008461	K7	
Surface No. 2		C = 0.0633393		R = 15.788
T = 7.70173	N = 1.5168	DEL N = 0.008054	BK7	
Surface No. 3		C = -0.0417607		R = -23.946
T = 5.7599	N = 1.6223	DEL N = 0.011707	SSK2	
Surface No. 4		C = -0.0222158		R = -45.0131
T = 53.5416	N = 1	DEL N = 0	Air	
Surface No. 5		C = -0.00973348		R = -102.738
T = -6.93917	N =-1	DEL N = 0	Air	
Surface No. 6		C = -0.00618526		R = -161.675
T = 98.65	N = 1	DEL N = 0	Air	

Image plane
Distance from exit pupil to image plane = 366.028

Canonical parameter – Forward orientation

		Module #1	Module #2	Module #3
TO	=	1.56047	1.44235	0.864916
T1	=	-0.0420836	0.0554986	-0.0608393
C1	=	5.76345	-4.33412	0.705474
C2	=	-2.02237	-2.30565	1.11018
TBAR	=	1.28091	1.12304	3.20915
TBPR	=	-0.185068	-0.334975	0.774232
E-W Class =		L(I)	L(II)	M(II)
H-W Class =		L(D)	L(A)	M(D)

Third-order aberration coefficients

SPH   = -6.02369E-07
COMA = -2.00407E-08
AST   = -1.26456E-08
PETZ = 0.00772572
DIST = -0.000356616

Chromatic aberration coefficients

LONG COL = 0.0016976
LAT COL = -6.60402E-08

---

in Fig.9.3 and Table 9.1. His work includes a study of the primary color aberrations as well as Petzval and distortion as functions of the three shape parameters.

Another possibility is in systems in which the role of image plane and pupil plane are interchanged such as condenser—projection lens combinations. Here the condenser images a source onto the entrance pupil of a projection lens whose focus lies on the exit pupil of the condenser. This sort of arrangement would involve an entirely different kind of coupling.

And that, Gentle Reader, concludes my tale. I must confess that I have no answer to the question "What good are modules?" I can only counter with Faraday's quip — "What good is a newborn babe!"

References

1.1 O.N. Stavroudis: J. Opt. Soc. Am. *57*, 741-748 (1967)
1.2 R.I. Mercado: Bol. Tonantzintla *2*, 317-326 (1978)
1.3 O.N. Stavroudis: J. Opt. Soc. Am. *59*, 288-293 (1969)
1.4 O.N. Stavroudis: "Two-Surface Optical Systems with Zero Third-Order Spherical Aberration"; OSC Tech. Rpt. No.37, University of Arizona (1969)
1.5 F.M. Powell: "y,ȳ Diagram Analysis of Two-Surface Optical Systems with Zero Third-Order Spherical Aberration"; Thesis, University of Arizona (1970). Also as OSC Tech. Rpt. No.55, University of Arizona (1970)
1.6 E. Delano: Appl. Opt. *2*, 1251-1256 (1963)
1.7 R.I. Mercado: "The Modular Principle in Optical Design"; Dissertation, University of Arizona (1973)
1.8 O.N. Stavroudis, R.I. Mercado: J. Opt. Soc. Am. *65*, 509-517 (1975)
1.9 D.W. Anderson: "The Practical Application of Modular Methods to Optical-System Design"; Dissertation, University of Arizona (1978)
1.10 D.W. Anderson: J. Opt. Soc. Am. *69*, 321-324 (1979)
1.11 W.T. Welford: *Aberrations of the Symmetrical Optical System* (Academic Press, London, New York, San Francisco 1974)
1.12 T. Smith: Proc. Phys. Soc. (London) *33*, 174-178 (1921)
1.13 W.T. Welford: Opt. Acta *15*, 621-623 (1968)
1.14 W. Brouwer: *Matrix Methods in Optical Instrument Design* (W.A. Benjamin, New York, Amsterdam 1972)
2.1 E. Delano: Appl. Opt. *2*, 1251-1256 (1963)
2.2 R.V. Shack: Proc. SPIE *39*, 127-140 (1973)
2.3 F.J. Lopez-Lopez: Proc. SPIE *39*, 151-164 (1973)
2.4 F.J. Lopez-Lopez: "The Application of the Delano y-ȳ Diagram to Optical Design"; Dissertation, University of Arizona (1973)
2.5 H.G. Zimmer: *Geometrical Optics* (Springer, New York, Heidelberg, Berlin 1970) pp.52-65
2.6 F.J. Lopez-Lopez: Appl. Opt. *9*, 2485-2488 (1970)
2.7 J. Kross: Optik *25*, 140-149 (1967)
2.8 H. Slevogt: Optik *30*, 431-436 (1970)
2.9 W. Besenmatter: Optik *47*, 153-166 (1977)
2.10 W. Besenmatter: Optik *47*, 381-390 (1977)
2.11 W. Besenmatter: Optik *48*, 289-304 (1977)
2.12 W. Besenmatter: Optik *49*, 1-15 (1977)
4.1 H. Cardano: *Artis magnae, sive de regulis algebraicis* (Paris 1545)
4.2 J. Eckmann: Supplement #7 Bull. Hist. Med., 59-67 (Johns Hopkins Press, Baltimore 1946)
4.3 O. Ore: *Cardano, the Gambling Scholar* (Princeton University Press, Princeton 1953)
4.4 D.E. Smith: *History of Mathematics* (Ginn, Boston 1925)
4.5 S. Gherardi: Arch. Math. Phys. *52*, 143, 488, 113-119, 143-147, 188 (1871)
4.6 N. Tartaglia: *Quesiti, et Inventioni Diverse* (Venice 1546)
4.7 V. Vilhelm: "Arithmetic and Algebra", in *Survey of Applicable Mathematics*, ed. by K. Rektorys (M.I.T. Press, Cambridge, MA 1969)

5.1 R.I. Mercado: Bol. Inst. Tonantzintla *1*, 265-276 (1975)
5.2 C.C. MacDuffee: *Theory of Equations* (John Wiley, New York 1954) pp.56-59
5.3 L.E. Dickson: *New First Course in the Theory of Equations* (John Wiley, New York 1939) pp.81-88
5.4 J.V. Uspenski: *Theory of Equations* (McGraw-Hill, New York 1948) pp.138-150
5.5 D. Anderson: "The Practical Applications of Modular Methods to Optical System Design"; Dissertation, University of Arizona (1978)
5.6 O. Ore: *Cardano, the Gambling Scholar* (Princeton University Press, Princeton, New Jersey 1953)
5.7 D.E. Smith: *History of Mathematics* (Ginn, Boston 1925)
5.8 V. Vilhelm: In "Arithmetic and Algebra", *Survey of Applicable Mathematics* ed. by K. Rektorys (M.I.T. Press, Cambridge, MA 1969)
6.1 R.I. Mercado: "The Modular Principle in Optical Design"; Dissertation, University of Arizona (1973)
6.2 O.N. Stavroudis, R.I. Mercado: J. Opt. Soc. Am. *65*, 509-517 (1975)
6.3 D.W. Anderson: "The Practical Application of Modular Methods to Optical System Design"; Dissertation, University of Arizona (1978)
6.4 F.M. Powell: "y-ȳ Diagram Analysis of Two-Surface Optical Systems with Zero Third-Order Spherical Aberration"; Thesis, University of Arizona (1974). [Also OSC Tech. Rpt. No.55, University of Arizona (1970)]
7.1 K. Schwarzschild: "Astronomische Beobachtungen mit elementaren Hilfs-mitteln", in *Neue Beiträge zur Frage des Mathematischen und Physika-lischen Unterrichts an den höheren Schulen*, ed. by C.F. Klein, C.V.E. Riecke (Teubner, Leipzig 1904)
7.2 A. Kohlschütter: "Die Bildfehler fünfter Ordnung optischer Systeme"; Dissertation, University of Göttingen (1908)
7.3 F. Wachendorf: Optik (Stuttgart) *5*, 80-122 (1959)
7.4 M. Herzberger: *Modern Geometrical Optics* (Interscience, New York 1958)
7.5 H. Buchdahl: *Optical Aberration Coefficients* (Oxford University Press, London 1954)
7.6 M. Rimmer: "Optical Aberration Coefficients", Appendix IV in *Ordeals II Program Manual*, ed. by M.B. Gold (Tropel, Fairport, NY 1965)
7.7 D.W. Anderson: "The Practical Application of Modular Methods of Optical System Design"; Dissertation, University of Arizona (1978)
7.8 D.W. Anderson: J. Opt. Soc. Am. *69*, 321-324 (1979)
8.1 R.I. Mercado: "The Modular Principles in Optical Design"; Dissertation, University of Arizona (1973)
8.2 D.W. Anderson: "The Practical Application of Modular Methods to Optical System Design"; Dissertation, University of Arizona (1978)
8.3 D.W. Anderson: J. Opt. Soc. Am. *69*, 321-324 (1979)
8.4 O.N. Stavroudis, R.I. Mercado: J. Opt. Soc. Am. *65*, 509-517 (1975)
8.5 R.I. Mercado: Bol. Inst. Tonantzintla *1*, 265-276 (1978)
8.6 R.I. Mercado, O.N. Stavroudis: J. Opt. Soc. Am. *65*, 1133-1140 (1975)
8.7 R.I. Mercado: Bol. Inst. Tonantzintla *2*, 317-326 (1978)
9.1 M. Herzberger: Opt. Acta *6*, 197-215 (1959)
9.2 R.E. Stephens: J. Opt. Soc. Am. *50*, 1016-1019 (1960)
9.3 R.E. Stephens: J. Opt. Soc. Am. *56*, 213-214 (1966)
9.4 R.I. Mercado: J. Opt. Soc. Am. *68*, 1451 (1978) Abstract

Subject Index

Aberrations *see* Third-order, Fifth-order, Seventh-order
Afocal system 2,13,183,240,252
  catadioptric 186
  hard-way coupled modules 111
  section on 155
Angle, semifield 106
Aperture *see* Pupil
Ars Magna 48,82
Astigmatism
  fifth-order 126-130
   canonical
    external 138
    extrinsic external backward 133
    extrinsic external forward 125
    extrinsic internal backward 131
    extrinsic internal forward 123
    internal 138
    intrinsic 120
    intrinsic backward 129
  example of zero 151,164
  extrinsic 117
  intrinsic 115
  lens design equation 140
  sagittal 115
  transverse 115
  third-order
   image 4

example of zero 151
in field curvature 41
in fifth-order extrinsic 130
general formula 11
module 55
for two surfaces 40
zero 56,58,96
pupil
  canonical 96,100
  general formula 12
  hard-way coupled 103
  lens-design equation 108,111
  for module 60
  for two surfaces 40
Axial chromatic aberration *see* Longitudinal chromatic aberration
Axial distance
  relative to module 33
  in y-y diagram 18-19

Back focal length 44
Back pupil plane 57
Backward orientation 4,98-100,126-133
  for fifth-order 126-133
   extrinsic internal 130-131
   extrinsic external 132-133
   intrinsic 129
  input-output 100
  optical parameters 98
  ray relationships 99
  for seventh-order 141-145

The Computer in Optical Research

Methods and Applications

Editor: B. R. Frieden
1980. 92 figures, 13 tables. XIII, 371 pages
(Topics in Applied Physics, Volume 41)
ISBN 3-540-10119-5

Contents: *B. R. Frieden:* Introduction. – *R. Barakat:* The
Calculation of Integrals Encountered in Optical Diffraction
Theory. – *B. R. Frieden:* Computational Methods of Proba-
bility and Statistics. – *A. K. Rigler, R. J. Pegis:* Optimization
Methods in Optics. – *L. Mertz:* Computers and Optical
Astronomy. – W. J. Dallas: Computer-Generated Holo-
grams.

H. J. Nussbaumer

Fast Fourier Transform and Convolution Algorithms

1981. 34 figures. X, 248 pages
(Springer Series in Information Sciences, Volume 2)
ISBN 3-540-10159-4

Contents: Introduction. – Elements of Number Theory
and Polynomial Algebra. – Fast Convolution Algorithms.
– The Fast Fourier Transform. – Linear Filtering Computation of Discrete Fourier Transforms. – Polynomial
Transforms. – Computation of Discrete Fourier Transforms by Polynomial Transforms. – Number Theoretic
Transforms. – References. – Subject Index.

Optical Data Processing

Applications

Editor: D. Casasent
1978. 170 figures, 2 tables. XIII, 286 pages
(Topics in Applied Physics, Volume 23)
ISBN 3-540-08453-3

Contents: D. Casasent:, H. J. Caulfield: Basic Concepts. –
B. J. Thompson: Optical Transforms and Coherent
Processing Systems-With Insights From Cristallography. –
P. S. Considine, R. A. Gonsalves: Oprätical Image Enhance-
ment and Image Restoration. –E. N. Leith: Synthetic Aperture Radar. – N. Balasubramanian: Optical Processing in
Photogrammetry. – N. Abramson: Nondestructive Testing
and Metrology. – H. J. Caulfield: Biomedical Applications
of Coherent Optics. – D. Casasent: Optical Signal Processing.

Springer-Verlag
Berlin
Heidelberg
New York

Optical Information Processing

Fundamentals

Editor: S. H. Lee
1981. 197 figures. XIII, 308 pages
(Topics in Applied Physics, Volume 48)
ISBN 3-540-10522-0

Contents: *S. H. Lee:* Basic Principles. – *S. H. Lee:* Coherent Optical Processing. – *W. T. Rhodes and A. A. Sawchuk:* Incoherent Optical Processing. – *G. R. Knight:* Interface Devices and Memory Materials. – *D. P. Casasent:* Hybrid Processors. – *J. W. Goodman:* Linear Space-Variant Optical Data Processing. – *S. H. Lee:* Nonlinear Optical Processing. – Subject Index.

B. Saleh

Photoelectron Statistics

With Applications to Spectroscopy and Optical Communication

1978. 85 figures, 8 tables. XV, 441 pages
(Springer Series in Optical Sciences, Volume 6)
ISBN 3-540-08295-6

Contents: Tools from Mathematical Statistics: Statistical Description of Random Variables and Stochastic Processes. Point Processes. – Theory: The Optical Field: A Stochastic Vector Field or, Classical Theory of Optical Coherence. Photoelectron Events: A Doubly Stochastic Poisson Process or Theory of Photoelectron Statistics. – Applications: Applications to Optical Communication. Applications to Spectroscopy.

From the reviews:
"...The material described in the book represents the work of many authors which have been widely scattered throughout the literature in various technical journals and books. Dr. Saleh has made a major contribution to the field by gathering most of the pertinent results in a single book and presenting the material concisely, correctly, and with a single consistent notation..."

Applied Optics

Springer-Verlag
Berlin
Heidelberg
NewYork

H. G. Zimmer

Geometrical Optics

Translator: R. N. Wilson
1970. 46 figures. VII. 171 pages
(Applied Physics and Engineering, Volume 9)
ISBN 3-540-04771-9